Tending Nature

Nathalie Gravel

Tending Nature

Fostering Eco-Citizenship in the Americas

Springer

Nathalie Gravel
Department of Geography
Université Laval
Quebec City, QC, Canada

ISBN 978-3-030-47061-6 ISBN 978-3-030-47062-3 (eBook)
https://doi.org/10.1007/978-3-030-47062-3

This Springer imprint is published by the registered company Springer Nature Switzerland AG
The registered company address is: Gewerbestrasse 11, 6330 Cham, Switzerland

To my daughters, Anne Salomé and Juliette,
You are the light of my life,
The guide of my actions,
The reason to keep hope,
The druid and elf that keep me on my toes,
My biggest fans and motivators.
Thank you for being you!

Preface

This book aims to provide a geographical framework for rethinking eco-citizenship based on bioregionalism principles, emphasizing the importance of community learning and participation in shaping public policy and normative frameworks for environmental protection.

It argues that eco-citizenship should be re-embedded in the lived regions of communities, where individuals can collectively define their "common good" and work towards environmental protection. The book examines successful Blue-Green-Bee conservation initiatives in the Americas, focusing on their collaborative management practices and environmental outcomes. It also discusses the concept of environmental citizenship, emphasizing the need for citizens to shape their duties and responsibilities based on their impacts on the environment and their sense of belonging to a place.

The book suggests that "bringing the state back in" as an actor of governance is crucial to accompany and support citizen-led initiatives and develop policies to protect the environment. It highlights the importance of institutionalizing change and fostering a shared environmental ethic through comprehensive environmental legislation at all levels of decision-making.

Overall, this book presents case studies from various countries in the Americas to demonstrate innovative approaches to Blue-Green-Bee conservation planning at municipal and regional scales and participatory natural resource management. It emphasizes the need for a deeper environmental awareness across society and highlights the role of public policies and municipal bylaws in promoting environmental sustainability to better face future challenges and enhance resilience. The book calls for a comprehensive approach to addressing multiscale environmental challenges, adapting programs to the diverse needs of communities and individuals, and shifting from a crisis-oriented discourse to one focused on creating transition landscapes.

Acknowledgments

The redaction of this book has been an extraordinary journey, involving collaboration with numerous individuals. I am deeply grateful for their contributions at various stages of the research, writing, editing, and map-making process. I also honor the memory of those who have passed, reassuring them that their legacy continues to live through me. Without their assistance and unwavering support, this book would not have come to fruition. While it's challenging to recall and acknowledge everyone who played a role in this endeavor, I extend my heartfelt thanks to the following:

- My family, including my two grown-up daughters, Anne Salomé and Juliette, my mother, Hélène, my late father, Raymond, who left us in 2020, and my daily fur companions, Bobby, the rescued American Cocker Spaniel and the late Kenny, a rescued Yorkipoo, who passed in 2021. Your constant support, daily assistance, and warm presence have meant the world to me.
- Louise Marcoux for the quality proof map-making from the Map Laboratory of the Department of Geography at Université Laval.
- Najat Bhiry, the former head of the Department of Geography at Université Laval, who played a crucial role in expediting the completion of the manuscript by lightening my workload.
- Former students from the Department of Geography at Université Laval: Émilie Sasseville, Josée Lemieux, and Griselda Tomé Hernández who have contributed to the research and Émilie Madore who has generously accepted to do the formatting of the manuscript.
- Individuals who generously shared their time, knowledge, and guidance throughout the research and writing process include: Dr. David Courtemanch—freshwater science and policy analyst at Nature Conservancy; Steve Brooke—the Kennebec Coalition project manager and officer for Trout Unlimited Kennebec Valley Chapter; Jeff Reardon—Maine Brook Trout Project Director for Trout Unlimited; Nick Bennett—former staff scientist and Healthy Waters director at the Natural Resources Council of Maine; Dr. Mark McCollough—biologist and wildlife artist from the Fish and Wildlife Service; Thom Whiting—Salmon river

activist, guide, and fisherman; Dr. Dan Kusnierz—researcher at the National Tribal Council; John Banks—former director of the Department of Natural Resources of the Penobscot Nation; Dr. Eric Venturini—Executive director of the Wild Blueberry Commission of Maine; Dr. Francis Drummond—professor Emeritus at the School of Biology and Ecology, University of Maine; Dr. Christine Feurt—head of the Saco Watershed Collaborative, Coastal Resilience scientist at the Rachel Carson National Wildlife Refuge, and research associate with the School of Marine and Environmental Programs at the University of New England in Biddeford; Danilo Bifone—president and founder of Muda MOOCA in São Paulo; Claudia Visoni—co-founder of the Horta das Corujas and a Green Party elect at the São Paulo state government; Gerson Pinheiro—founder of S.O.S. Abelhas sem Ferrão in São Paulo; Dr. Erika Sagae—Graduated Ph.D. Student at Universidade Federal de Santa Catarina, Brazil; Dr. Clécio Azevedo da Silva; Dr. Pedro Jacobi, Professor at the Institute of Energy and Environment at the University of São Paulo; Malu Gandra—former director of the São Paulo Biosphere Reserve; Dr. Edgar F. de Luca—former director of the São Paulo Forest Institute (IF); Dr. Lídice Almeida Paraguassu—professor at the Federal Institute of Bahia; Sergio Shigeeda—the late founder of the Horta da Saúde and environmental activist in the City of São Paulo; and, lastly, Hiroyassu Uehara—late researcher from the São Paulo Forest Institute (IF) who shared an office with me and many insights.

- Those who shared valuable materials with me: Benoît Gauthier, Steve Brooke, Juliette Richard, and the Monmouth Museum in Maine, with special mention to the rewilding Maine artist and garden designer, Kdb Dominguez who generously accepted to share her work of art for the cover of this book.
- Friends I encountered on various paths in Quebec, Canada, Maine, Ontario, Mexico, Cuba, and Brazil among whomArtur Cezar Santinello—director of the Forest Museum of São Paulo; Jean Denis Brisson—president of the Friends of the Van den Hende Garden, Quebec City; Anthony Cohutt, owner of Antiques USA in Arundel, Maine, Christopher Whiting, Philip Thibault—executive assistant at the Association of Canadian Geographers, and Peter Sharp Vargas whose moral support made the final edits easier.
- Dr. Paola Jofre for accepting to proofread my chapters.
- Drs. Sidclay Pereira Cordeiro and Hector José Martinez Arboleya for being examiners of the manuscript.
- Dr. Thomas Klak, who hosted me as a visiting professor during my 2018–2019 research leave at the Department of Environmental Sciences at the University of New England in Biddeford, and Dr. Barry Smit from the Department of Geography, Environment, and Geomatics at the University of Guelph, who hosted me during my previous research leave (2010–2011), and
- To all my former and current graduate and undergraduate students of the Geography and Agroforestry programs at Université Laval for their questions and the shining light in their eyes.

I deeply appreciate their invaluable contributions to this work. I appreciate my undergraduate students for their eagerness to learn about conserving nature in their own backyard and getting involved in environmental groups to become engaged citizens. The conversations we've had in class over the last 20 years have been inspiring and have fueled my commitment to continue. This book is another stone in the bridge I am building to get us to the other side of the river. No matter how wide the river is and how long it takes to build the bridge, never ever give up! The Earth will cry when we give up.

I would also like to express my gratitude to my two co-authors, Griselda Tomé Hernández and Sonia Emilia Silva Gómez, for their valuable contributions to the writing of Chap. 9. Despite the collaboration of many individuals, I take full responsibility for any potential errors.

Funding

The author expresses gratitude to the Government of Canada for funding the research in Brazil (Chap. 8) through the Canada-Brazil Joint Research Project on Environmental Governance from 2014 to 2016. Ethics approval has been obtained, when necessary, for individual chapters.

About This Book

This book aims to explore the complex theme of achieving harmony between the natural environment and human society by reframing the concept of ecological citizenship at the local scale of foodsheds and watersheds. Showcasing an innovative Blue-Green-Bee Framework to help understand the complex mesh of intertwining ecological dimensions of tending Nature, the chapters bring out successful experiences of policies and initiatives from various countries in the Americas. These initiatives span diverse spheres, including forest, river, agricultural, and urban ecosystems, as well as land-use planning for the conservation of wild pollinator habitats. The work underscores the urgency of comprehensive environmental education programs and sustained outreach endeavors to cultivate ecological awareness and responsible behaviors among citizens.

This book highlights the significance of implementing community governance of the commons, reinforced by multi-level action plans to translate Blue, Green, and Bee conservation public policies into tangible, sustainable actions. Furthermore, it underscores the importance of community-based learning and citizen involvement as integral aspects of participatory governance, with the goal of addressing the prevailing environmental inequities in our modern world.

Divided into two parts with a total of ten chapters, this book provides a geographical framework for the re-evaluation of eco-citizenship, drawing inspiration from the principles of bioregionalism. After establishing the need to reintegrate eco-citizenship within the local communities' lived spaces and presenting the analytical framework based on the foodshed/watershed scale (Chaps. 1 and 2), the first part explores "The Strength of Eco-Networking" across four chapters. By showcasing case studies from Canada (Chaps. 3 and 5), Costa Rica (Chap. 4), and the United States (Chap. 6), this section anchors participatory governance in the formation of alliances, either horizontally with a diverse range of stakeholders or multiscalarly, linking local initiatives with regional and national partners who share similar aspirations for environmental preservation.

The second part of the book, titled "The Role of Social Innovation and Public Policies in Designing Transition Landscapes," comprises three chapters. It is strategically written to engage policymakers, encouraging active involvement in sustainable territorial planning. Drawing on fieldwork conducted in the United States and Cuba (Chap. 7), Brazil (Chap. 8), and Mexico (Chap. 9), this section explores the concept of "transition landscapes." It underscores the urgency of comprehensively safeguarding both natural and urban ecosystems, not only for human interests but also for non-human entities, guided by principles of care and compassion. This transition presents a pragmatic approach to moving humanity away from crisis scenarios and toward a more comfortable zone where the quality and diversity of natural assets are guaranteed as a heritage for future generations.

The work also offers a comprehensive examination of the concept of environmental citizenship. It underscores the profound interconnectedness between humans and the broader ecosystem, emphasizing the necessity of defining the civic commons that communities aspire to protect and the responsibilities that align with the environmental consequences of stakeholders' actions, whether within or beyond their national borders. The case is made for the revitalized role of the state, particularly at the municipal and regional levels, as an active participant in environmental governance, fostering collaboration with other societal actors rather than seeking to preserve decision-making dominance. This collaborative approach aims to shape policies that prioritize the well-being of both humans and the integrity of ecosystems.

In conclusion, the book emphasizes the necessity of developing comprehensive environmental legislation to protect citizen-led initiatives from the influence of powerful interests and to formalize transformative actions and landscapes. It underscores the importance of adapting Blue, Green and Bee conservation measures to be inclusive of marginalized groups and tailoring programs to involve them as contributors in sustainable planning. Furthermore, it highlights how political and economic factors can significantly impact civil society initiatives and make their plans vulnerable. In the context of planning smart cities and Blue-Green-Bee programs, it stresses the critical importance of strategic multiscalar alliances between stakeholders and media members, multidisciplinary training for individuals engaged in environmental governance, and a legislative capacity that empowers them to adapt programs to the diverse needs and perspectives of various communities while countering subversive moves from detractors.

The main objective of this book is to enhance the comprehension of the essential elements and ethical aspects related to citizen involvement in environmental protection. This goal will be achieved through a comprehensive examination of cases that demonstrate successful collaborative environmental governance throughout the Americas.

Some of the most important specific goals include:

- Contribute to the theoretical discourse on ecological citizenship, participatory environmental governance, and sustainable management of the commons by examining various case studies and re-evaluating existing literature from a fresh perspective.
- Highlight the potential of eco-networking facilitated by non-governmental actors and their influence on transformative actions at municipal, regional, and national levels.

- Investigate the challenges and opportunities for environmental conservation and protection in the Americas, exploring the dynamics of actors who support successful environmental stewardship.
- Document and share successful stories of citizen participation and civil society collaborations that have played significant roles in the decision-making process, leading to socio-ecological transformations in the Americas.
- Provide insights for public policymakers on how to promote behavioral change and move toward the co-management of public natural resources and transition landscapes.
- Introduce the Blue-Green-Bee framework as a foundational approach for reevaluating projects and public policies within the context of bioregions, serving as an operational guide for future knowledge generation and action plans.

Contents

List of Figures

List of Tables

About the Author

Nathalie Gravel, Ph.D. is a geographer deeply committed to environmental sustainability. She earned her Ph.D. from Université Laval in Quebec City, Canada, in 2003, and later completed a post-doctoral fellowship at Yale University's Agrarian Studies program and Council for Latin American Studies and the Caribbean. Since 2005, she has held the position of associate professor at Laval University, where she teaches courses on rural and food geography, Latin America, globalization, and chairs the Agroforestry graduate programs.

Dr. Gravel's research is focused on participatory natural resources governance, food security, rural studies, and regenerative agriculture, leading her to present at numerous conferences and publish scientific papers. Proficient in French, English, Spanish, and Portuguese, she served as the President of the Canadian Association of Geographers (2022–2024) and actively participates in various boards and associations.

In 2009, Dr. Gravel authored the textbook *Geography of Latin America: A Culture of Uncertainty* (French: *Géographie de l'Amérique latine. Une culture de l'incertitude*), published by Université du Québec Press (PUQ). She is dedicated to fostering inclusive community engagement, promoting geography, and addressing global challenges through education and research.

Contributors

Griselda Tomé Hernández, Ph.D. is trained as an environmental engineer, with a Master's degree in Agricultural Development Strategies, a Ph.D. in Geographical Sciences from Université Laval, Quebec City, Canada, and postdoctoral studies at the Colegio de Postgraduados. She has 10 years of experience in issues related to territorial transformations, water management, and governance at the watershed, sub-watershed, and local levels. She is author and co-author of various articles in scientific journals and books, and speaker at forums and conferences in different cities in Mexico, Canada, France, and Brazil. She has been working as a water management consultant for the past 9 years in the states of Baja California, Veracruz, Oaxaca, Tlaxcala, and Puebla. Additionally, since 2020, she has held the position of professor-researcher at the Instituto Tecnológico Superior de San Martín Texmelucan (ITSSMT).

Sonia Emilia Silva Gómez, Ph.D. graduated with a Bachelor's degree in Social Anthropology from Universidad Veracruzana in 1983. In 1993, she achieved a Master's in Strategies for Regional Agricultural Development (EDAR) from Colegio de Postgraduados en Ciencias Agrícolas (ColPos). Later, in 2001, she completed her Ph.D. in EDAR. Her research is predominantly centered on the sustainability of rural-urban systems, environmental pollution, and social risk. She has assumed leadership and collaborative roles in various research projects covering diverse themes, including native corn varieties, rural housing, water management in rural areas, risks posed to populations by environmental contamination, peasant knowledge, and ecotourism. She currently holds the position of professor researcher at the Department of Sustainable Development, Scientific Institute of the Benemérita Universidad Autónoma de Puebla (BUAP). She is recognized at National Researchers System Level I, has a PRODEP Profile, and is listed in the BUAP Researchers Registry. Additionally, she has authored and co-authored numerous scientific articles and books.

Abbreviations

APA	Environmental Protection Area (Brazil)
ASOTUR	Association of Tourist Guides in the Irazú Volcano Region
BC	Branch Committee
CATIE	Tropical Agricultural Research and Higher Education Center
CBVCT	Central Talamanca Volcanic Biological Corridor
CC	Central Committee
CEPAGRO	Center for the Study and Promotion of Group Agriculture
Cobri-SURAC	Reventado Agua Caliente Biological Corridor
CONAGUA	Mexican National Water Commission
CONAPO	Consejo Nacional de Población
CSA	Community-Supported Agriculture
DEP	Department of Environmental Protection
DIA	Diversity, Integration, and Autonomy
EIA	Energy Intensive Agriculture
ENGO	Environmental Non-Governmental Organization
EPA	Environmental Protection Agency
ESD	Education for Sustainable Development
FERC	Federal Energy Regulatory Commission
GEF	Global Environment Facility
GRCA	Grand River Conservation Authority
GTA	Greater Toronto Area
IMFN	International Model Forest Network
KRAC	Kennebec River Valley Anglers' Coalition
MF	Model Forest
MINAET	Ministry of Environment, Energy, and Telecommunications of Costa Rica
MOFGA	Maine Organic Farmers and Gardeners Association
MPAP	Multistakeholder Policy Formulation and Action Planning
MRSP	Metropolitan Region of São Paulo
MUDA	Urban Agroecology Movement
MZO	Special Zoning Order

NASCO	North Atlantic Salmon Conservation Organization
PAA	Food Acquisition Program
PANC	*Planta Alimentícia Não Convencional* (Unconventional Food Plant)
PES	Payments for Ecosystem Services
PLANAPO	Plan for Agroecology and Organic Production
PMAU	Municipal Urban Agriculture Program
PNAPO	National Policy of Agroecology and Organic Production
PPD	*Programa de Pequeñas Donaciones* (Small Grants Program)
PR plants	Pollen-Rich plants
PRO-AURP	Urban and Peri-Urban Agriculture Program
PRONAF	Program to Strengthen Family Farming
PT	*Partido dos Trabalhadores* (Workers' Party)
RMF	Reventazón Model Forest
SESC	Community Service Center
SGP	Small Grants Program
SINAC	*Sistema Nacional de Áreas de Conservación* (National System of Conservation Areas)
SPSS®	Statistical Package for Social Sciences
SRT	Social Regulation Theory
UA	Urban Agriculture
UNPD	United Nations Development Program
UPA	Urban and Peri-urban Agriculture
USDA	U.S. Department of Agriculture
WWW	Wellington Water Watchers

Chapter 1
Introduction

Abstract Landowners grapple with extensive dilemmas in managing private properties, from tree cutting to riverbank protection, amid an industrialized landscape reshaped by irreversible environmental impacts. Despite nature's regenerative capacity, human intervention remains crucial in safeguarding ecological integrity amid escalating environmental crises.

Recent shifts in stakeholder awareness toward sustainable practices reflect growing commitments driven by reputational concerns and intergenerational responsibilities. This introduction explores cultural contexts shaping environmental stewardship and societal attitudes toward ethical resource usage, laying the groundwork for the book's exploration of transformative social alliances and innovative policy integrations.

This chapter advances the book's thesis on Nature–Human reconciliation through ecological citizenship and participatory governance, emphasizing grassroots initiatives. It underscores successful eco-networks and innovative policy integrations as pathways toward sustainable landscape transitions, guided by inclusive conservation efforts and sustainable development goals.

Keywords Environmental sustainability · Industrialization impacts · Human · Nature connection · Stakeholder awareness · Sustainable practices · Environmental stewardship · Ethical resource usage · Social alliances · Policymaking · Ecological citizenship · Participatory governance · Grassroots initiatives · Eco-networks · Sustainable transitions

1.1 Introductory Remarks

The dilemmas faced by landowners in managing nature on private lots can be extensive, ranging from decisions about tree cutting, septic tank installation, protection of riverbanks, to the maintenance of flowering vegetation, weeding, and dealing with perceived nuisances like bugs. Often, humans forget their inherent connection to nature, viewing it as something to be manipulated rather than embraced as a part of

N. Gravel, *Tending Nature*, https://doi.org/10.1007/978-3-030-47062-3_1

themselves. Industrialization has provided the means to reshape landscapes dramatically, altering rivers, deforesting, and clearing riverbanks with sometimes irreversible consequences. Once environmental degradation reaches a certain point, reversing the trend becomes challenging. Land can be so profoundly impacted that it resembles a total war on nature.

In agricultural fields, unsustainable practices have led to the disappearance of soil organisms, fauna, and insects in and above the ground. Despite some displayed willingness, reversing these trends can be difficult, requiring time and patience to see initial results. Eventually, with the right conditions, nature can begin to heal itself. We've witnessed the recovery of endangered species like the bald eagle in the United States, which was removed from the endangered list following changes in the paper-making industry and the removal of dams, allowing the resurgence of fisheries and free-flowing rivers (see Chap. 6). Despite these notable achievements, nature still requires human assistance.

Humans, who have significantly altered the course of nature since the Anthropocene, must now act as stewards to defend and protect nature's rights for the benefit of both the natural world and future generations. The urgency to save nature's bounty is critical, as humanity stands at the precipice. The participation of all stakeholders is essential for such a collective endeavor.

In recent years, a noticeable shift in awareness has emerged among various stakeholder groups, with many expressing a surprising interest in changing their practices. This transformation often stems from the realization that their reputations were at stake, and the recognition that their children would bear the consequences. In his work "Spaces of Hope," David Harvey advocates for a shift in how we relate to work and the environment, expressing faith in human qualities to achieve a utopian dream grounded in the realization of socio-environmental justice. This book takes a step further by showcasing victories of social groups and alliances that have reversed trends in unexpected ways.

The wisdom embedded in the alliances detailed in the following chapters empowers individuals, rekindling the light that, for many, has flickered amid eco-anxiety, particularly among the youth. While Elinor Ostrom has paved the way for understanding the logic of civil society groups in managing public natural resources (the commons) and caring for their local environment successfully, the book goes beyond the dynamics of decision-making and organizing. It delves into the historical and cultural context of local residents and other actors, examining the discourses they produce, their interactions with higher governmental hierarchies, and the coexistence of diverse knowledge, from scientific expertise to locally based know-how and belief systems.

The relationship that humans have developed with common natural resources, those not owned by anyone but by all, is complex. It's easy to transfer the responsibility of managing them to others and not feel concerned about their future. However, as Ostrom emphasized, we will collectively suffer the consequences while blaming others for abusing the resources more than we do. The perception, care, and protection of common resources raise important questions. How do we use them when no one is looking? Do we refrain from using or abusing these resources out of fear of punishment from social regulations or authorities, or do we have a self-censoring mechanism preventing us from taking advantage of resources that don't belong to us?

Studying human behavior when faced with free resources is insightful. Do people delve into them to take advantage of free items, or do they consider leaving some for others, or even refuse to take any at all? For example, when foraging in a forest for edibles, do individuals serve themselves to the point of leaving nothing behind, or do they respect the plants and trees that offer their bounties, allowing the vegetation to survive and reproduce? These questions provide valuable insights into human attitudes toward common resources and the ethics that guide their use.

Do we leave enough of Nature for the next person to visit, or, by extension, for the next generations to enjoy? Unfortunately, many individuals leave marks of their presence in natural settings, such as cigarette butts, empty bottles, and various types of garbage. This behavior raises the question of why people leave traces of their visit when the natural amenities are the primary reasons for their exploration. Introducing outside materials to nature contributes to diminishing these amenities, making places less attractive to visitors and even to those who left the litter the next time they return. This highlights the importance of responsible and sustainable practices when interacting with natural environments to preserve their beauty and ecological integrity for future generations.

As a child, I participated in a class experiment where I picked up trash on my way to school, and the amount I collected left a lasting impression. Throughout my life, I have continued to clean up public and private spaces, including streets, riversides, beaches, and parks. During several summers, I served as a park guardian in an urban riparian park along the St. Lawrence River in the province of Quebec, Canada. However, this role wasn't always conducive to making friends, as people often dislike being reprimanded for improper behavior. This sentiment was particularly evident during the pandemic, where parks became havens for freedom, but at the expense of the surrounding environment.

While visitors craved freedom, frustrations over social distancing and restrictions led to rule violations. Unleashed dogs, littering, failure to clean up after pets, building unattended fires, and leaving ashes on the river shore became common occurrences. Despite clear rules posted at trail entrances and associated fines for violations, some acted as if environmental education had never begun, reminiscent of the 1970s.

My approach as a park guardian evolved from confrontation to education over time. I discovered that the most effective way to address issues was by asking questions, attempting to spark an interest in the park rather than confronting individuals. While it was heartbreaking to witness unleashed dogs trampling endangered marsh plants that the foundation worked hard to protect, I realized that those who enjoyed nature needed just a little encouragement to become better stewards. To kindle this spark, especially among younger generations, I started creating videos on social media.

The park provided proper infrastructure for garbage disposal, yet individuals emptied ashtrays in parking lots near bins and disregarded designated disposal areas. Park managers were continually astonished by the lack of consideration. The pandemic raised questions about its impact on society's relationship with nature and common resources. Did it make us more careless or disobedient in public behavior? These observations prompt reflection on whether my experiences have colored my perception or if there are genuine shifts in public conduct that environmental stewards must address. I hope for the latter, acknowledging any gains and lessons learned by those dedicated to caring for nature.

Continuing my commitment, I still find myself picking up cigarette butts from the road in front of my house every other day. This routine aims to counteract the pollution they cause to the nearby lake, a fact that many might not be aware of—just one cigarette butt can contaminate more than 500 liters of water (Rath et al., 2012). Despite being small, cigarette butts, like microplastic beads, have a lasting and profound impact on nature. These filters, found in coastal areas and cities during cleanup campaigns, constitute 40% of all litter. They take about 12 years to disintegrate and release approximately 4000 chemical compounds, including 50 known carcinogens.

The global distribution of smokers exacerbates the issue, making cigarette filters a pervasive problem for oceans and coastal lands. Rath et al. (2012) identified the belief that cigarette butts are not litter as the primary cause of this pollution. Responsible smokers would keep the butts in their pockets until finding an appropriate place for disposal. In my effort to pick up discarded butts, I hope to raise awareness among my neighbors. This simple act often sparks interesting conversations and, at times, prompts expressions of guilt from those who have littered.

Over time, I've come to realize that comprehending the perceptions individuals form throughout their lives regarding nature and their surrounding environments is crucial for developing an effective environmental education strategy. These perceptions are dynamic and can be subject to change, with education serving as a catalyst for such transformations.

1.2 Tending Nature and Common Resources in a Nutshell

This book delves into the journey toward Nature–Human reconciliation, employing an ecological citizenship approach that places strong emphasis on participatory governance. It underscores the vital role of involving state actors in the formulation of Blue-Green-Bee conservation policies, as well as the significance of community learning and citizen engagement in the pursuit of environmental justice.

In recent years, there has been a discernible diminishment in federal funding for environmental protection initiatives across the Americas, driven largely by the interests of industrial capitalism. This has engendered a discernible void in environmental protection efforts, necessitating the emergence of grassroots and citizen-led green initiatives, albeit in a somewhat disorganized and inclusivity-challenged manner. Consequently, a substantial portion of the responsibility for environmental conservation now falls upon individuals operating within the private sphere.

The first part of the book, titled "The Strength of Eco-Networking," comprises four chapters. Chapter 3 presents two Ontario-based case studies of successful multiscalar citizen-led struggles to protect farmland and natural ecosystems on the outskirts and within the Greater Toronto Area against aggregate extraction industries and urbanization pressures. Chapter 4 focuses on actor mapping and negotiation related to natural resources stewardship within a Model Forest in Costa Rica. Chapter 5 explores the architects of blue cities in Guelph, Canada, who draw knowledge and inspiration from the Canadian water network and contribute applied knowledge to it in return. Chapter 6 delves into multipartite civic alliances in the State of Maine, United States, dedicated to supporting not only the rights of Atlantic salmon but also the recovery of entire river ecosystems following dam removals.

These coalitions have proven to be powerful, securing significant victories that resonate beyond their respective territories, serving as inspiring examples for others. The engaged actors within these alliances have made a substantial impact on the stewardship of Nature, reinforcing the idea that decision-makers at multiple levels can champion the cause of Nature and develop public policies aligned with their values and principles.

Furthermore, the diversity of representation within these alliances has brought together individuals from various backgrounds and interests, highlighting the importance of selecting causes that can unite a diverse cross-section of society. Striving for consensus, at times, holds greater significance than achieving specific objectives, as it leads to positive societal transformation.

The second part of the book, titled "The Role of Social Innovation and Public Policies in Designing Transition Landscapes," outlines innovative approaches that have the potential to be incorporated into public policies at various governance levels, ranging from municipal to national. The central aim is to shift away from crisis-ridden landscapes and offer a vision of hope. Hope becomes attainable when communities transition from emergency and crisis modes to strategic planning for their territories, guided by common objectives and sustainable development goals.

Chapter 7, which explores the historical evolution of farmsteads in New England, presents a model for transitioning away from crisis-ridden agricultural landscapes toward more diverse and pollen-rich plant reservoirs. Alongside the pre-modern New England homestead, Cuban knowledge and agroecological experiments serve as a significant source of inspiration for future biodiverse agricultural landscapes. While the decline of honeybees remains a pressing issue, native bees represent an essential backup plan to ensure a diverse diet. Policymakers need to recognize that part of the solution lies in the rewilding of both urban and periurban landscapes, as well as diversified agricultural areas.

Chapter 8 illustrates how Brazilian urban and periurban agriculture has become an integral part of green infrastructures, providing ecological services such as creating havens for pollinators. Driven by citizens' advocacy and supported by a comprehensive framework of laws, bylaws and urban agriculture policies, Brazilian metropolises have incorporated elements such as biological corridors, linear parks, and urban community gardens to enhance and green urban ecosystems. These resulting transition landscapes hold the promise of a more abundant, resilient, and optimistic future.

Finally, Chap. 9, authored by Griselda Tomé Hernández and co-authored by myself and Sonia Emilia Silva Gómez, unveils a successful social innovation in drinking water distribution in the State of Puebla, Mexico. Drawing inspiration from historical water management practices, the community of San Gabriel Chilac, situated near the pre-Columbian Tehuacán Dam, operates a citizen-managed system that ensures equitable and frequent distribution of drinking water in a semiarid region. The chapter provides insights into their self-regulated system, showcasing how residents achieve this feat independently, with only minimal municipal intervention. This system, reminiscent of Elinor Ostrom's work on Spanish water-managing communities, demonstrates the enduring effectiveness of trust and equitable sharing of responsibilities.

References

Rath, J. M., Rubenstein, R. A., Curry, L. E., Shank, S. E., & Cartwright, J. C. (2012). Cigarette litter: smokers' attitudes and behaviors. *Int J Environ Res Public Health. 9(6), 2189–203.* https://doi.org/10.3390/ijerph9062189. Accessed August 28, 2024.

Chapter 2
Revealing the Conceptual Framework

Abstract This chapter introduces the Blue-Green-Bee Framework, a comprehensive approach to guide future actions in an environmentally conscious world. The framework integrates three core elements: Blue (representing water and aquatic ecosystems), Green (symbolizing terrestrial ecosystems and green spaces), and Bee (highlighting pollinators' biodiversity and interconnected ecological services). It emphasizes their interdependence and aims to inform public policies for environmental sustainability, conservation, and biodiversity protection.

This chapter outlines the foundations and processes necessary for developing public policies that foster citizen engagement, empowering individuals to actively contribute to the shaping and stewardship of their valued territories. It highlights the ongoing need for environmental education to foster greater collaboration between citizens and Nature. Drawing from years of teaching geography, the author advocates for legal reforms that empower citizen groups to shape policies aligned with sustainable development principles for the common good.

Keywords Agroecological transition · Biodiversity protection · Bioregionalism · Blue-Green-Bee Framework · Citizen actions · Common good · Ecological citizenship · Ecological networking · Environmental governance · Environmental sustainability · Managing the commons · Participatory governance · Public policies · Sustainable development · Water and energy soft paths

2.1 Introduction

To accomplish the afore-mentioned objectives, the conceptual framework supporting this book will be concisely delineated. The primary foundations include theories and perspectives on environmental governance, ecological citizenship, ecological networking, managing the commons, bioregionalism, water and energy soft paths, and agroecological transition.

N. Gravel, *Tending Nature*, https://doi.org/10.1007/978-3-030-47062-3_2

Drawing on the insights gained globally from five decades of environmental education,[1] a paradigm shift is required—moving away from the Nature–Culture dichotomy toward an integrated perspective that positions humans within Nature. The imperative to reassess our detrimental interactions with nature and replace them with nondestructive practices is emphasized to address the prevailing environmental crisis. The book advocates for a Blue-Green-Bee conservation commitment to nurture natural spaces and restore fully functioning ecosystems. The Blue-Green-Bee matrix represents the living environment that sustains human life and activities, as well as all life on this planet. Respecting the principles that safeguard water, bees, and forests in human actions can set Planet Earth on the path to recovery from the Anthropocene aggressions, signifying the period post-1950 when humans began exerting a significant and profound impact on Nature, transforming landscapes and geology (Ernstson & Swyngedouw, 2019). The term Anthropocene denotes a geological period unique in its human-induced shift, distinguishing it from previous natural geological periods (Idem).

Practical actions to safeguard pollinators biodiversity at the citizen's level will be outlined, applicable in various settings, including backyards, parks, and living, working, and leisure spaces. Moreover, the book will provide recommendations for public policy decision-making, serving as guidelines for decision-makers at different levels of government, NGOs, and those working in both small and large municipalities.

2.2 The Building Blocks of Participatory Commons Governance

What is intended by building blocks are operational and methodological units of analysis that allow us to put in practice and render intelligible the theoretical and conceptual framework: the bioregion, foodshed, and watershed are three of those units that we will need to help us link theory and praxis and hopefully suggest creative ways to craft public policies that encourage an ecological transition.

Social innovations often emerge in response to the imperative of managing scarce or finite resources, particularly when government-led initiatives have proven ineffective. Nobel laureate Elinor Ostrom defines commons as "common-pool resources," referring to goods that can be depleted if overused but are challenging to exclude people from using (Nordman, 2021: 2). Ostrom's work underscores the significance of self-regulation and communal systems in handling finite common-pool resources, challenging the notion that "communities are condemned to the fate of overusing their commons" (Ibid). Through numerous instances of successful community management, she questioned Garrett Hardin's "Tragedy of the

[1] This is the case if we take the First United Nations Conference on Environmental Education in Tbilisi in 1972 as a benchmark.

Commons" and formulated principles for collective decision-making regarding shared resources. Several case studies, including those explored in this book, align with Elinor Ostrom's research on community-based resource management.

One notable example is Törbel, a Swiss alpine community, where pastoralists devised a system for equitable water sharing crucial for their cattle and overall survival in the arid alpine landscape. Mc C. Netting's 1972 ethnographic account of Törbel, revisited by Ostrom (1990, republished, 2010), illustrates the community's efforts to balance individual interests with commons management.

Nestled between 1500 and 2000 meters of altitude in the Swiss Alps, Mc C. Netting's 1981 book titled "Balancing on an Alp. Ecological change and continuity in a Swiss mountain community" illuminates the unique challenges and opportunities of sustenance in this alpine region. Cultivation practices include rye, potatoes, vineyards, and fruit trees. The area boasts the driest climate in Switzerland, attributed to the Foehn wind, which imparts a Mediterranean climate, echoing as far as Italy. An exploration of out-migrations reveals their pivotal role in maintaining a steady flow of resources and income into local agricultural production. The author places emphasis on examining the strategic choices made by individual peasants, offering insights into their significance within the broader context of alpine land-use.

Ostrom examined communal water management practices of the Törbel villagers, as detailed in Mc C. Netting's narrative, showcasing their harmonious coexistence with the landscape, described as follows: "The landscape of Törbel is an extension of its people, and the possibilities and limitations of their society are in part reflections of their habitat" (Mc C. Netting, 1981: 5).

> The village territory that I have crudely sketched is above all a cultural landscape…The terrain has been used and modelled for generations by a resident human population. Without the grazing of their livestock and the mowing of the hay, those expansive clearings that so delight the eye would be as forested as the foothills of the Rockies or the Cascades (Idem: 3).

> The local environment is what people have to work with, and their tools and traditions define the means at their disposal for this task (Idem: 5).

> Given limited resources and equally fixed means of utilizing them, individual players of the peasant subsistence game may take several courses of action. We may term these strategies intensification, expansion, and regulation […] Perhaps the most pressing problem and the one that demanded specifically social solutions was that of regulation. How was access to the available resources to be allocated? (Mc C. Netting, 1981: 43).

I read "Balancing on an Alp" very early in my study years, and its influence has been lasting. It prompted me to contemplate the specificities of island-type territories disposing of finite natural resources, such as cultural macrocosms (Idem: 42). The challenges of managing finite common-pool assets are numerous, but if they are anchored in a definite territory, it is easier to grasp the consequences of mismanagement and lacks such as accounted for in Hardin's Tragedy of the Commons in 1968 (Nordman, 2021). Many closed societies have successfully evolved around self-sustenance, a concept that may find resonance in the bioregional agenda proposed in the following pages.

Climate change, along with the globalization of risks and the boomerang repercussions of human actions and mismanagement of the natural resources (Beck, 2001), as exemplified by events like the avian flu, underscores the urgency of attempting to govern the Earth's commons from the bottom up (Griffon, 2013, 2017). This entails adopting a glocal[2] perspective, initiating efforts at the local level and relying on the ripple effect for broader impact.

However, the frameworks presented by Ostrom and Mc C. Netting may not always be directly applicable in open territories with multiple stakeholders managing fluid common natural resources, such as air, freshwater, oceans, climate, and soils, that may be shared by all or by many groups or nations. This book centers on citizens' actions, often organized within civil society networks, negotiating with industrial, governmental, and transnational entities for better environmental futures. While Ostrom's collective management principles offer guidance in numerous scenarios, it is essential to incorporate a theoretical framework influenced by the human geography of power and critical social theory to truly account for a multilevel and multistakeholder governance of environmental resources. This becomes particularly crucial for comprehending the dynamics involved in decentralized governance that engages local populations in shared management power arrangements. The term "participatory governance" is used to characterize this dynamic interaction among stakeholders.

2.3 Participatory Environmental Governance and Sustainability

One of the focal concepts of this book is participatory environmental governance. Environmental governance entails the fair exertion of power in managing natural resources within a specific territorial entity, aiming to shape the planning of resource utilization according to the objectives of various actor groups within the entity, achieving consensus. In governance studies, an actor-mapping approach has evolved to help examine each interest group asserting access to the relevant natural resources (Gill & Reed, 1999; Pierce & Dale, 1999; Klooster, 2006; Jean, 2000).

In prior works, we characterized governance as "an emerging order within a system that arises from social interactions and serves as both the outcome of these interactions and the means by which actors can act and interpret" (Gravel & Lavoie, 2009: 13). This definition emphasizes the role of power and the involvement of groups of actors with differing interests (stakeholders), distinguishing it from the purely technocratic or business-oriented management of resources, akin to a planning exercise devoid of negotiations among actors.

[2] The concept of a "glocal approach" involves considering both global and local factors in decision-making and problem-solving. It recognizes the interdependence of global and local elements, emphasizing the need to address issues with consideration for both scales.

Coupled with autonomy and fostering the establishment of a network of actors independent of established power structures, the concept of participatory governance is intricately connected to the idea of social capital (Ikerd, 2005; Bebbington, 2004). Social capital, as defined by Enjolras (2005), pertains to the level of organization and readiness for collective action. This readiness is contingent on the quality of the actor network, their capacity for participation, and the involvement of social economy organizations.

In this definition, participatory environmental governance transcends mere administrative actions. Instead, it involves establishing arenas for participation and dialogue among diverse stakeholders addressing a shared issue. Within this framework, participants collaboratively define objectives, roles, and responsibilities, devise and adopt solutions, ultimately reaching a consensus to implement the agreed-upon measures and establish a continuous review process in a context of perpetual evolution (Enjolras, 2005). This conceptualization interprets governance in the environmental context as:

> The sum of the many ways in which people and institutions, public and private, manage their affairs in common. It is a continuous process through which conflicting or divergent interests can be sorted out and cooperative actions can be taken. This includes formal institutions and regimes with enforcement powers, as well as informal arrangements that people and institutions have agreed to or perceive to be in their interests (Commission on Global Governance, 1995: 2).

Sustainable development rests on four foundational pillars, each contributing to the overall resilience and balance of societal progress. These pillars are:

- Economic Sustainability: This pillar emphasizes the necessity of economic systems that are resilient, inclusive, and do not deplete resources. It involves fostering economic growth while ensuring long-term stability and equity.
- Social Sustainability: Social sustainability focuses on creating inclusive societies that prioritize social equity, justice, and well-being. It entails addressing issues such as poverty, education, healthcare, and social cohesion.
- Environmental Sustainability: At the core of sustainable development, this pillar underscores the importance of environmental conservation and responsible resource management. It aims to protect ecosystems, biodiversity, and natural resources for current and future generations.
- Governance: The capacity to steer a system or entity toward achieving consensual objectives while maintaining balance among the other three pillars. The practice of governance entails safeguarding cultural heritage, diversity, and identity, and nurturing cultural behaviors that align with equity, the environment, and enhance social welfare.

These four pillars form an interconnected framework, and sustainable development requires a holistic approach that considers the interdependence of economic, social, environmental, and cultural factors. In the framework of sustainable development, this consensual space of governance is illustrated as the fourth circle around the diagram representing the pillars of this concept, that is, society, economy, and environment (see Fig. 2.1).

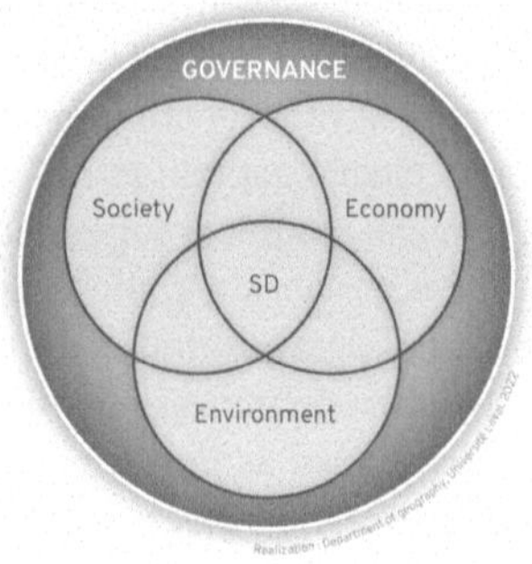

Fig. 2.1 The four pillars of sustainable development. (Inspired by the NGO Vision Development Durable (Gauthier, 2007))

The overarching aim of participatory environmental governance at the community level is to foster cohesive communities engaged in collaborative projects, equipped to address their challenges through negotiation, solidarity, and a shared vision (Gagnon, 2006). This vision is in harmony with Ostrom's eight principles for collective resource management and resonates with bioregionalist concepts, which prioritize sustainability (Sale, 1996; Ostrom, 2010; Nordman, 2021). Sustainability, in this context, is understood as the balancing act of socioeconomic interactions that enable the self-regeneration of natural resources and align with the carrying capacity of a given environment. It is not inherently tied to traditional notions of development, as argued by Ikerd (2005) and Sachs (2003), who critique the juxtaposition of sustainability and development.

2.4 Ecological Citizenship

Environmental citizenship emerges from a proactive inclination to actively contribute to the well-being of one's immediate natural surroundings and landscape. It aligns with civic engagement and a commitment to safeguarding the integrity of one's territory, reflecting a form of civic identity on a multi-scale terrain. The concept of territory, viewed as a space evoking a sense of belonging and a motivation to protect and nurture, is integral to the evolution of environmental citizenship. This term garnered meaning as I sought to understand the dynamics underlying successful civic struggles. Emphasizing the significance of these civic triumphs is crucial, given their rarity and as a response to the environmental concerns of younger generations. While numerous civic victories are attainable in the global North, the proposed actor mapping and case studies offer potential utility in inspiring members of civil society in other countries to mobilize and defend their territories when they perceive the opportune moment.

Echoing the wisdom of my late mentor, political scientist and anthropologist Dr. James C. Scott at Yale University, as articulated in his work "The Moral Economy of the Peasant" (1976), prioritizing security considerations should precede the decision to confront state-led or private sector initiatives. Prudent strategic considerations counsel the wisdom of timing, as articulated by Nobel Peace Prize laureate Rigoberta Menchu (1992), who cautioned "If you think you are going to have your head cut, then you do not show it" (translated from Spanish, Burgos, 2007). She regretted that for too long, Indigenous people of her country, Guatemala, needed to remain silent to preserve their lives (Idem).

It is imperative from the outset to recognize the divergent degrees of agency that civil societies possess across different nations, marked by variations in influence, resources, support networks, and certainties, affecting their ability to assert and defend their rights successfully. Opposing stakeholders can be either foreign entities, each characterized by distinct natures and anticipated impacts, or domestic actors such as known businesses or the state itself, championing development in pristine territories. Numerous instances abound of protracted struggles, some concluded, others ongoing. Notable among them is the Northeastern Mainers' battle against river dams owned by national utility companies and a Toronto-based Canadian electricity company, a topic further expounded upon in Chap. 6.

The decline in biodiversity has been associated with a decrease in cultural diversity, influenced by global homogenizing trends and values where humans are each time more distant from nature. UNESCO's Education for Sustainable Development (ESD) advocates for a renewed commitment to environmental stewardship, encouraging individuals to take responsibility for their habitat's quality.

Environmental stewardship, influenced by the perspectives of communities closely connected to their immediate surroundings, highlights principles of accountability, unity, and environmental citizenship. This approach reflects a commitment to responsibly manage and care for the natural environment, fostering a sense of shared responsibility and promoting active engagement in environmental matters. This concept aligns with a holistic worldview, reminiscent of indigenous thinking, where individuals naturally tend to their land and care for their habitat because they see themselves as part of Nature, rather than separate entities. Environmental citizenship, as defined in this context, involves active citizen participation in advancing sustainability, challenging traditional notions of citizenship to address environmental issues. It plays a crucial role in the shift toward governance, emphasizing collective action in environmental policy and politics (Carlsen, 2004; Sale, 1996; Kassam, 2009; Pallett, 2017).

Key elements of eco-citizenship include:

Connection to Place Eco-citizens feel a deep connection to their local environment and land, recognizing them as places of belonging and identity. This connection to the land, which involves emotions, can be mobilized to encourage social engagement in the immediate region.

Stewardship Eco-citizens act as stewards of the land and water, taking responsibility for the well-being of local ecosystems. This may involve sustainable practices, conservation efforts, and advocating for policies that promote environmental health.

Sustainability Eco-citizens prioritize sustainable living practices, considering the environmental impact of their choices. This includes mindful consumption, waste reduction, and supporting initiatives that promote ecological balance.

Advocacy Engaged eco-citizens often become advocates for environmental causes. They may participate in community initiatives, support policies for environmental protection, and contribute to efforts aimed at addressing ecological challenges. Acknowledging larger global problems require selfless motivations, often stemming from a general interest in protecting life on the planet and ensuring the continuity of the human experience.

Holistic Vision Eco-citizenship fosters a comprehensive perspective on development, challenging conventional approaches that may harm the environment. It advocates for integrated solutions that acknowledge the interconnectedness of social, economic, and environmental systems, exemplified by the circular economy, which prioritizes resource sharing and reuse over purchasing new items. The holistic vision of ecosystems encompasses consideration for all living habitats, including underground (living soils), terrestrial, water, and the various atmospheric layers, all interacting harmoniously and supporting the interconnected flow of biological diversity.

Calls for more decentralization of power, requiring local participation in resource-allocation processes, and respect of indigenous knowledge and human and environmental rights are being observed globally, in all countries. Local groups advocating for the preservation of green belts, opposition to pristine forest clear-cutting on the Canada and US Pacific coast, activism in support of salmon rivers throughout the United States, and indigenous groups worldwide advocating for cultural diversity and land rights have sparked a vibrant networking activity. This collaboration involves NGOs, churches, and indigenous movements, fostering a collective effort to address environmental and cultural concerns. A common message is being heard: Human communities are adapting and designing new strategies for sharing information, combining indigenous and scientific knowledge, and putting up resistance through the affirmation of their cultural and biological diversity while offering alternatives to the capital-accumulation dominant worldview aimed at homogenizing and reducing diversity (Kassam, 2009).

In response to global corporate activities and pressures felt at all geographical scales (from micro to global) and to governments' weakness in the face of these pressures, the widespread tentacles of civil society can be seen as a product of site-specific socioeconomic interactions and tensions. Given the importance of context to understanding issues of sustainability, the networking capacity of civil society must be studied within particular contexts, by means of case studies. Civil society groups need to be considered among other social actors to understand which needs human communities try to satisfy.

In summary, eco-citizenship reflects a proactive and conscious approach to living that prioritizes environmental sustainability, conservation, and a sense of responsibility for the health of the planet. Public participation is essential for the localized application of sustainable development principles (Gauthier & Gagnon, 2013; Pallett, 2017).

2.5 Eco-networking

The increasing recognition of the importance of preserving cultural and biological diversity has led multilateral organizations, including UN agencies like UNESCO, UNEP, FAO, and UNICEF, to provide the conceptual framework for grassroots development and educational initiatives. These organizations rely on grassroots organizations for outreach, legitimizing the existence of NGOs. However, this has also imposed new expectations on NGOs to find creative solutions to complex regional problems and enhance understanding of sustainability issues. The subsequent discussion explores the role of civil society networks in promoting local participation in ecosystem management. However, a prerequisite for community leaders to engage in environmental management is being well-informed and trained in environmental education and sustainable development to effectively communicate and represent citizens' opinions and requests.

Civil society organizations' inclination to form alliances with regional, national, and international entities can be seen as a means to enhance their legitimacy beyond the local context and garner attention from politicians (Carrier & Jean, 2000). The development of a stronger identity and increased visibility are strategic approaches employed by organizations to establish themselves as significant political actors.

Eco-networking involves the establishment of connections and collaborations among various actors within the environmental and sustainability domain. These actors can include individuals, organizations, institutions, and other entities that share a common interest in promoting ecological well-being and sustainable practices. The term "alliances" refers to the partnerships and cooperative efforts formed among these actors to address environmental challenges and work toward shared goals.

In the context of eco-networking alliances, actors may encompass a wide range of participants, such as environmental activists, community groups, nonprofit organizations, businesses, government agencies, researchers, indigenous peoples, and other stakeholders (Gravel and Koné, 2017). These alliances aim to create a network of interconnected entities that can pool their resources, knowledge, and efforts to achieve more significant and positive impacts on environmental conservation and sustainability. Through collaborative initiatives and coordinated actions, eco-networking alliances contribute to fostering a collective and effective approach to address pressing environmental issues and promote a healthier planet.

2.6 Bioregionalism

The concept of a bioregion refers to a "life-place" or encompassing macrocosm that defines the territory regularly traversed by an individual and shares common natural and cultural attributes at a regional scale. Originating as a response to the adverse impacts of development-associated growth, bioregionalism, as proposed by Sale (1992) and Stuppia (2022), played a crucial role in zero-growth policies in municipalities like Petaluma, California. It also influenced a community movement in the Mattole River valley, where local solutions were sought for issues such as the reintroduction of salmon.

Aligned with the deep ecology movement, bioregionalism views humans as integral parts of the natural environment, urging them to align their lifestyles with ecological limits. This philosophy advocates for decentralization, local engagement, and participation in regional politics, gaining traction in the United States and resurging in Europe around the 2000s. Sale (1996) proposed localized consumption and waste management to reduce waste production and encourage innovation. The bioregional movement gained momentum in the 1970s and 1980s, influencing international conferences like the 1972 Stockholm UN Conference on the Environment. The publication of Sale's "Dwellers in the Land" in 1985 marked the integration of bioregional ideas into regional politics, initiating mainstream acceptance.

Simultaneously, a counter discourse advocating globalization and the complementary nature of exchanges, as per Ricardo's theory, accompanied the expansive growth of global capitalism (Desrochers & Shimizu, 2012). The unfolding anthropic and natural impacts of such choices have underscored the limitations of international trade, subcontracting, and consumption-based lifestyles, prompting a reevaluation of everyday actions to minimize ecological footprints. This perspective, adopting a glocal approach, emphasizes the significance of even small actions and recognizes every piece of land as a potential biodiversity reservoir.

Marris (2011), in "Rambunctious Gardens: Saving Nature in a Post-Wild World," contends that National parks, guided by a conservationist approach, concentrate on safeguarding highly diverse and exceptional ecosystems. On the opposite end of the spectrum is the protection of everyday spaces like backyards, roadside ditches, airports, and highway marginal lands. Here, the regular citizen assumes the role of an agent of change, collaborating with other stakeholders in the shared bioregion or community. This is the message I choose to convey to my students, urging them to recognize their responsibility and influence in reversing potentially harmful trends. It is a deliberate strategy to prevent succumbing to eco-anxiety, recognizing that fear can paralyze action and hinder positive change.

2.7 Foodshed

Bioregionalists at the origins of the watershed inspired the neologism "Foodshed" (Kloppenburg et al., 1996). Recognizing one's belonging to a local foodshed or watershed brings a sense of "connection and responsibility" (Idem). Connection helps one feel grounded in a territory and identify with the landscape, its history, and culture. Responsibility because one becomes answerable for his/her actions that have an imprint on the "shed." In an imaginative extension akin to the watershed, permaculturist Arthur Getz has coined the term "foodshed" to encourage thoughtful consideration of the origins and transportation of our food. The "foodshed" is not merely a metaphor; much like its counterpart, the watershed, it functions as a conceptual and methodological unit of analysis, offering both a framework for contemplation and a guide for practical action (Idem).

Those related concepts are useful to design targeted environmental education and awareness campaigns and as tools of public policies, which is the reason why an emphasis is placed on them in this book. Not only are these concepts important to call upon to delineate the working territory—this is the first of Ostrom's (2010) recommendation for a working environmental participatory governance—but also to muster citizens recalling their responsibilities as part of an interrelated regional system. What does that mean in practical terms? This means that if natural resources are mismanaged, wasted, or polluted, due to those interrelations, the effects will be felt by the inhabitants of the water or foodshed. The residence and workplace define in a good part one's consuming spaces and therefore the role in the possible foodsheds (maybe more than one).

Without those clear spatial boundaries, actors may easily deny their responsibilities, pointing to other actors that may have a greater impact on a natural resource than they do. In a watershed, what differentiates the residents who live alongside a lake, on the other side of the street from the lake, or further away in the village? What is the "allowed" pesticides for each one to apply on their lawns? Are there differences according to the distance from the lake? No, in all cases, pesticides will leach in the ground with the rain and will flow toward the lake or river. Only those directly on a lakeshore or riverbanks would have a more direct impact because there is less vegetation and soil filtering the lawn pesticides before reaching the body of water (this is why a minimum 3 m-wide (9′8″) riparian buffer zone is supposed to be left with vegetation such as trees, bushes, and tall grass in countries such as Canada). Overall, the bonds that tie the residents to their territory are the same, and the responsibilities regarding the protection of natural resources or food resources (including soils) are the same.

Philip Ackerman-Leist (2013), based in Vermont, advocates for "rebuilding the foodshed," as reflected in one of his books' titles. He is a passionate supporter of farmers' markets, viewing them as a means to bring us closer to the local foodshed.

2.8 The Agroecological Transition

The decline in life within agricultural soils is a concern that, while less tangible for urbanites who may not engage in gardening, is no less critical than the decline in pollinators. Biodiversity narratives often emphasize the visible aspects of the ground and above-ground levels, overlooking the essential but unseen life in the soils. The emergence of the worldwide Living Soils movement[3] reflects the need to raise awareness about soil life for a successful ecological transition.

Shifting the focus to soils allows us to recognize their vital roles not only in food production but also in construction, housing, infrastructure, and green spaces. Problems such as erosion threatening communities often stem from inadequate soil protection, planning, or a failure to respect the limits set by Mother Nature. The current challenges posed by climate change underscore our lack of wisdom and lead to crisis landscapes globally, characterized by events like erosion, landslides, wildfires, floods, or droughts. Many of these threats share a common cause: our vulnerability and inconsiderate exploitation of the Earth over the last 70 years, coinciding with the Anthropocene (Griffon, 2013). The question arises: What does it take to transform these crisis landscapes into transition landscapes?

The unrestricted consumption patterns of humans come with trade-offs, involving externalities that aren't factored into the cost of food and often go unnoticed by consumers. Questions about the water usage in meat production, the distance traveled to deliver products, or the exploitation of underage children in crop harvesting often remain unaddressed. While social and economic considerations might be more accessible motivators for some to change their consumption habits, the environmental motivations, particularly in connection to food production, appear to be less well understood.

Agroecology is a holistic and interdisciplinary approach to agriculture that integrates ecological principles into farming systems. It emphasizes the creation of sustainable and resilient agricultural ecosystems by considering the interactions between plants, animals, humans, and the environment. Agroecology seeks to optimize the use of local resources, reduce reliance on external inputs like synthetic fertilizers and pesticides, and enhance the overall well-being of farming communities (Olivier, 2021).

Key principles of agroecology include biodiversity promotion, conservation of natural resources, recycling of organic matter, and the enhancement of ecological processes to support agricultural productivity. This approach aims to create farming systems that are environmentally sound, economically viable, and socially just.

[3] The Living Soils movement is a global initiative that encompasses efforts and practices worldwide that promote sustainable soil management, regenerative agriculture, and the enhancement of soil health. The movement emphasizes the importance of soil as a living and dynamic ecosystem essential for ecological balance and sustainable food production. Various organizations, farmers, and communities across different countries actively participate in the Living Soils movement to promote practices that prioritize soil conservation and regeneration.

Agroecology is often considered a response to the environmental and social challenges posed by conventional industrial agriculture.

Altieri 9, often regarded as the father of agroecology, asserts that the primary cause of ecological disorders in conventional agriculture is the monocrop system (In Funes-Monzote, 2010). Various botanical and entomological studies, such as those by Girard et al. (2012) and Koh et al. (2016), identify row crop systems as being responsible for the loss of native pollinator habitats (see Chap. 7).

The agroecological approach celebrates the interconnections between all natural elements on the farm, including humans, and emphasizes the retrieval of the natural self-regulating and healing potential of ecosystems (Bourguignon and Bourguignon, 2017). In a manner similar to how First Nations view a river as embedded in its entire ecosystem, comprising earth, water, and atmosphere along with associated wildlife (see Chap. 6), the agroecologist farmer perceives these ribbons of life as conduits for energy flow—from the soil to the trees, to the crops below, and back to the land—nurturing the diverse living beings it hosts, including animals, insects, birds, and human beings (Griffon, 2013).

2.9 The Water and Energy Soft Paths

The soft path approach, first articulated by Amory B. Lovins in his 1976 publication "Energy Strategy: The Road Not Taken?" in Foreign Affairs and further elaborated in his 1977 book "Soft Energy Paths: Toward a Durable Peace" to address the energy crisis of the 1970s, underwent a substantial expansion to include water conservation efforts three decades later (Holtz & Brooks, 2011: 39). Canadian water researchers Oliver M. Brandes and David B. Brooks, founders of the POLIS Project on water sustainability, recognized the parallels between Lovins' work in energy and the demand management practices in water. Lovins made a notable contribution by acknowledging the presence of numerous small-scale, frequently local, alternative energy sources. Simultaneously, he emphasized the importance of reducing demand through improved demand management. This shift in perspective highlighted the significance of regarding energy as a service rather than an ultimate goal (Gravel and Koné, 2017; Brooks et al., 2011).

Central to the soft path approach is the conceptualization of water as a service rather than a finite resource, with a key emphasis on ecological sustainability. This approach prioritizes matching water quality to specific end-use needs and adopts a future-oriented planning strategy, working backward to address present challenges. Distinguished from water demand management, the soft path approach takes into account carrying capacity considerations and strives for the long-term attainment of collective objectives (Brandes & Brooks, 2006, 2007).

The soft path approach advocates planning based on demand rather than solely relying on resource availability. This entails setting goals for a 10-, 25-, or 30-year horizon tailored to a specific territorial entity. It emphasizes the importance of reducing demands for water resources and nonrenewable energy to reterritorialize our lives within a bioregion and align our needs with its carrying capacity. In contrast, the hard path prioritizes substantial infrastructure investments over demand reduction. Advocates for ecological transition believe that addressing future challenges requires a shift in consumption habits, embracing reuse, applying circular economy principles, and exploring the untapped potential of the green economy (Brooks et al., 2009).

These two approaches will be a central focus in numerous chapters throughout this book.

2.10 Crafting Public Policies Anchored in the Blue-Green-Bee Framework

The guiding framework proposed in this book for shaping future actions in a caring and solidarity-driven world revolves around the principles of Blue, Green, and Bee conservation. The *Blue-Green-Bee Framework* is a holistic approach that integrates three essential elements: Blue (representing water and aquatic ecosystems), Green (symbolizing terrestrial ecosystems and green spaces), and Bee (highlighting biodiversity, pollinators, and interconnected ecological systems). This framework emphasizes the interconnectedness of these elements and aims to guide the design of public policies that address environmental sustainability, conservation, and biodiversity protection. It encourages a comprehensive and balanced perspective, considering the intricate relationships between water resources, terrestrial ecosystems, and the vital role of pollinators like bees in maintaining ecological health.

This book aspires to catalyze the development of public policies that encourage and support citizen actions, fostering their active participation in decision-making processes for the territories that hold significance to them. The Rio+20 international assembly, marking two decades since the Earth Summit, underscored a shared commitment among states and municipalities to prioritize the agenda for the implementation of local participatory governance (Gravel, 2018; Gagnon, 2006). While sustainable practices and policies have found a place in the governance structures of many regions and municipalities, the extent of co-management and public participation remains diverse. There is a continued need for environmental education to permeate through various societal layers, fostering greater collaboration between citizens and nature. Public policies should further endorse and support initiatives by eco-citizens, empowering them to shape local environments in alignment with their desires and visions.

Over my two decades of teaching geography at the university level, I've come to realize that genuine change involves modifying laws and legal systems. In this regard, I align with Ostrom's belief that groups of citizens should possess the power to alter existing laws and regulations and propose new ones that better align with sustainable development principles for the commons.

References

Ackerman-Leist, P. (2013). *Rebuilding the foodshed: How to create local, sustainable, and secure food systems*. Chelsea Green Publishing.

Altieri, M. A. (1989). Agroecology: A new research and development paradigm for world agriculture. *Agriculture, Ecosystems & Environment, 27*(1–4), 37–46. Accessed August 26, 2024. https://doi.org/10.1016/0167-8809(89)90070-4

Bebbington, A. (2004). Livelihood transitions, place transformations: Grounding globalization and modernity. In R. N. Gwynne & C. Kay (Eds.), *Latin America transformed. Globalization and Modernity* (2nd ed., pp. 173–192). Edward Arnold.

Beck, U. (2001). La politique dans la société du risque. Revue du. *MAUSS, 17*, 376–392. Retrieved at: https://doi.org/10.3917/rdm.017.0376. Accessed 18 Jan 2024.

Bourguignon, L., & Bourguignon, C. (2017). *Manifeste pour une agriculture durable*. Actes Sud.

Brandes, O. M., & Brooks, D. B. (2006, revised in 2007). *The soft path for water in a nutshell* (Victoria, B. C., POLIS Project). University of Victoria and Friends of the Earth Canada. Retrieved at: https://dspace.library.uvic.ca/bitstream/handle/1828/7963/The%20Soft%20Path%20for%20Water%20in%20a%20Nutshell.pdf?sequence=1&isAllowed=y. Accessed 18 Jan 2024.

Brooks, D. B., O. M. Brandes and S. Gurman (2009, revised in 2011). *Making the most of the water we have* (The Soft Path Approach to Water Management). Earthscan.

Burgos, E. (2007). *Me llamo Rigoberta Menchú y así me nació la conciencia* (20th ed.). Siglo Veintiuno Editores.

Carlsen, L. (2004). *Conservation or privatization? Biodiversity, the global market and the Mesoamerican Biological Corridor*. In G. Otero (Ed.), *Mexico in Transition. Neoliberal globalism, the state and civil society* (pp. 52–71). Zed Books.

Carrier, M., & Jean, B. (2000). La reconstruction de la légitimité des collectivités rurales. Entre gouvernement et gouvernance. In M. Carrier & S. Côté (Eds.), *Gouvernance et territoires ruraux. Éléments d'un débat sur la responsabilité du développement* (pp. 41–63). Presses de l'Université du Québec.

Commission on Global Governance. (1995). *Our global neighborhood*. The Report of the Commission. Retrieved at: http://www.gdrc.org/u-gov/global-neighbourhood/. Accessed 18 Jan 2024.

Desrochers, P., & Shimizu, H. (2012). *The Locavore's Dilemma: In Praise of the 10,000-mile Diet*. Public Affairs.

Enjolras, B. (2005). Économie sociale et solidaire et régimes de gouvernance. *Revue internationale de l'économie sociale, 296*, 59–69.

Ernstson, H., & Swyngedouw, E. (Eds.). (2019). *Urban political ecology in the Anthropo-obscene: Interruptions and possibilities*. Routledge.

Funes-Monzote, F. R. (2010). Cuba: A national-level experiment in conversion. In S. R. Gliessman & M. Rosemeyer (Eds.), *The conversion to sustainable agriculture. Principles, processes, and practices* (pp. 205–238). CRC Press.

Gagnon, C. (2006). L'agenda 21 local : un outil de développement durable et viable, sous-utilisé par les collectivités territoriales québécoises. In M. Robitaille, J.-F. Simard, & G. Chiasson (dir), *L'Outaouais au carrefour des modèles de développement* (pp. 133–143). CRDT, CRDC, UQO.

Gauthier, M., & Gagnon, L. (2013). *La participation du public dans les démarches municipales de développement durable: Principes, conditions de réussite, enjeux et dispositifs* (Les cahiers du CRGRNT, no 1307). Université du Québec en Outaouais.

Gill, A., & Reed, M. (1999). Incorporating postproductivist values into sustainable community processes. In J. T. Pierce & A. Dale (Eds.), *Communities, development, and sustainability across Canada* (pp. 166–189). UBC Press.

Girard, M., Chagnon, M., & Fournier, V. (2012). Pollen diversity collected by honeybees in the vicinity of Vaccinium spp. crops and its importance for colony development. Pollination biology research in Canada: Perspectives on a mutualism at different scales. *Botany, 90*(7), 545–555.

Gravel, N. (2018). *La Contrée en montagnes, 10 ans de gouvernance territoriale partagée*. Paper presented at the International Geographic Union Conference in partnership with the Canadian Association of Geographers, Quebec City, August 6–10.

Gravel, N. (2024). Antecedentes y trayectoria de la gobernanza participativa de los recursos naturales: lecciones de Canadá, Costa Rica y Brasil [Antecedents and trajectories of natural ressources participative gouvernance: Lessons from Canada, Costa Rica and Brazil. In G. Massardier and V. Ugalde (eds) Politicas Publicas: de las teorias a las metodologias, Mexico, Colegio de México, pp. 471–512. ISBN 978-607-564-562-9. https://isbnmexico.indautor.cerlalc.org/catalogo.php?mode=detalle&nt=408824

Gravel, N. (ed.) (2009). Environmental Governance in Latin America. *Special Issue of the Canadian Journal of Latin American and Caribbean Studies*, 34(68). Guest editor. Conference proceedings. http://www.jstor.org/stable/i40084749

Gravel, N., & Koné, A. (2017). The Guelph Water Connection: The Contribution of Actor-Network Theory (ANT) to the Study of Water Management in Guelph, Ontario. *Les Cahiers de géographie du Québec. Special issue on water in Canada and Brazil*. Ed. Louis Guay and Ana Britto. *61*(174), 489–512.

Gravel, N., & Lavoie, A. (2009). Introduction. Environmental Governance in perspective. *Canadian Journal of Latin American and Caribbean Studies. Special issue on environmental governance in Latin America, 34*(68), 5–18. Retrieved at: https://www.jstor.org/stable/i40084749. Accessed 18 Jan 2024.

Griffon, M. (2013). *Pour des agricultures écologiquement intensives*. Nouvelles éditions de l'Aube.

Griffon, M. (2017). *Écologie intensive : la nature comme inspiration pour l'agriculture et la société*. Buchet-Chastel.

Holtz, S., & Brooks, D. B. (2011). In the beginning: Soft energy paths. In D. B. Brooks, O. M. Brandes, & S. Gurman (2009, revised in 2011). *Making the most of the water we have. The soft path approach to water management* (pp. 37–49). Earthscan.

Ikerd, J. (2005). *Sustainable capitalism. A matter of common sense*. Kumarian Press.

Jean, B. (2000). Une nouvelle gouvernance territoriale pour accompagner la restructuration rurale. In M. Carrier y S. Côté (eds) *Gouvernance et territoires ruraux* (pp. 223–244). Presses de l'Université du Québec.

Kassam, K.-A. S. (2009). *Biocultural diversity and indigenous ways of knowing. Human ecology in the Arctic*. University of Calgary Press/The Arctic Institute of North America.

Klooster, D. (2006). Environmental certification of forests in Mexico: The political ecology of a nongovernmental market intervention. *Annals of the Association of American Geographers, 96*(3), 541–565.

Kloppenburg, J. J., Hendrikson, J., & Stevenson, G. W. (1996). Coming in to the foodshed. *Agriculture and Human Values, 13*, 33–42. Retrieved at: https://link.springer.com/article/10.1007/BF01538225. Accessed 18 Jan 2024.

Koh, I., Lonsdorf, E. V., Williams, N. M., Brittain, C., Isaacs, R., Gibbs, J., & Ricketts, T. H. (2016). Modeling the status, trends and impacts of wild bee abundance in the United States. *Proceedings of the National Academy of Sciences, 1*(113), 140–145. https://doi.org/10.1073/pnas.1517685113. Accessed 13 Jan 2024.

Lovins, A. B. (1977). *Soft energy paths: Toward a durable peace*. Ballinger Publishing and Friends of the Earth International.

Marris, E. (2011). *Rambunctious garden: Saving nature in a post-wild world*. Bloomsbury.

Mc C. Netting, R. (1981). *Balancing on an Alp: Ecological Change and Continuity in a Swiss Mountain Community*. Cambridge: Cambridge University Press.

Nordman, E. (2021). *The uncommon knowledge of Elinor Ostrom. Essential Lessons for Collective Action*. Island Press.

Olivier, A. (2021). *La révolution agroécologique. Nourrir tous les humains sans détruire la planète*. Écosociété.

Ostrom, E. (1990, 2010) *Gouvernance des biens communs. Pour une nouvelle approche des ressources naturelles*. Bruxelles, De Boeck Supérieur. Col. « Planète en jeu ».

Pallett, H. (2017). Environmental citizenship. In M. Goodchild, A. Kobayashi, W. Liu, D. Marston, D. Richardson, & N. Castree (Eds.), *The international encyclopedia of geography: People, the earth, environment, and technology*. Wiley Blackwell. https://doi.org/10.1002/9781118786352.wbieg0474. Accessed 18 Jan 2024.

Pierce, J. T., & Dale, A. (Eds.). (1999). *Communities, development, and sustainability across Canada*. UBC Press.

Sachs, W. (2003). *The development dictionary: A guide to knowledge as power*. Zed Books Ltd.

Sale, K. (1985). *Dwellers in the land. The bioregional vision*. Sierra Club.

Sale, K. (1996). Principles of bioregionalism. In J. Mander & E. Goldsmith (Eds.), *The case against the global economy and for a turn toward the local* (pp. 471–484). Taylor and Francis.

Stuppia, P. (2022). Reinhabiting Earth. 50 years of Bioregional Utopias. *Développement durable et territoires*. Special issue on Territories during the Pandemic/Experimentations of Ecological Transitions. 13(2). Retrieved at: https://doi.org/10.4000/developpementdurable.21572. Accessed 18 Jan 2024.

Website

Gauthier, B. (2007). *Vision développement durable*. http://visiondeveloppementdurable.com/index.php?option=com_content&view=article&id=68&Itemid=68. Accessed 18 Jan 2024.

Part I
The Strength of Eco-Networking

Chapter 3
Navigating Pressures on the Ontario Greenbelt

Abstract Environmental citizenship intertwines with civic engagement, embodying a commitment to preserve the integrity of one's territory, establishing a form of civic identity across multiple scales. Recent challenges in Ontario, Canada, including attempts to establish a quarry in prime farmlands for extractive purposes and the endeavor to lift protection from the Duffins Rouge Agricultural Preserve, the Greater Toronto Area Greenbelt, and the Oak Ridges Moraine Conservation Area, were met with robust opposition from informed and engaged citizens. These initiatives sparked concerns about their potential impact on endangered species, headwaters protection, and nearby First Nations communities. Environmental NGOs and empowered citizens effectively thwarted both projects, compelling provincial leaders to prioritize the safeguarding of shared resources. While acknowledging these citizens' triumphs, it is imperative to explore alternative strategies and glean insights from other successful initiatives, recognizing the continued vulnerability of protected lands in the Greater Toronto Area.

Keywords Agglomerates extraction · Civic engagement · Duffins Rouge Agricultural Preserve · Endangered species · Environmental citizenship · Environmental NGOs · Farmland protection · First Nations communities · Greater Toronto Area · Greenbelt · Headwaters · Oak Ridges Moraine · Ontario, Canada · Protection · Quarry

3.1 Introduction

In this chapter, we direct our focus to a dual triumph achieved by the residents of Southern Ontario in the Great Lakes region, Canada. The victories pertain to the thwarting of the Melancthon Mega-quarry project in 2011 and, more recently, the preservation of the Ontario Greenbelt's protection that was at risk due to residential development. The initial success was a collective effort against the Highland

N. Gravel, *Tending Nature*, https://doi.org/10.1007/978-3-030-47062-3_3

Companies, supported by a Boston hedge fund. The proposed mega-quarry aimed to exploit the province's valuable potato land for aggregate extraction to supply the cement industry and facilitate the construction of additional residences, parking lots, and highways (Suzuki, 2012). The second victory faced a threat from the provincial head of government and collusion practices favoring real estate developers. The Ontario General Auditor played a crucial role in compelling the reversal of the decision in September 2023.

During my sabbatical leave in 2010–2011 at the University of Guelph, Ontario, my awareness of the Melancthon Mega-quarry project was indirectly kindled by the proliferation of lawn signs throughout the city posted by its spirited opponents, prompting me to delve deeper into the issue.

3.2 Case Study No. 1: The Melancthon Mega-Quarry Battle

Situated in Dufferin County, northwest of the Greater Toronto Area (see Fig. 3.1) and the World Heritage Niagara Escarpment, the township of Melancthon occupies a crucial position with dolostone soils. Positioned atop the recharge zone for the

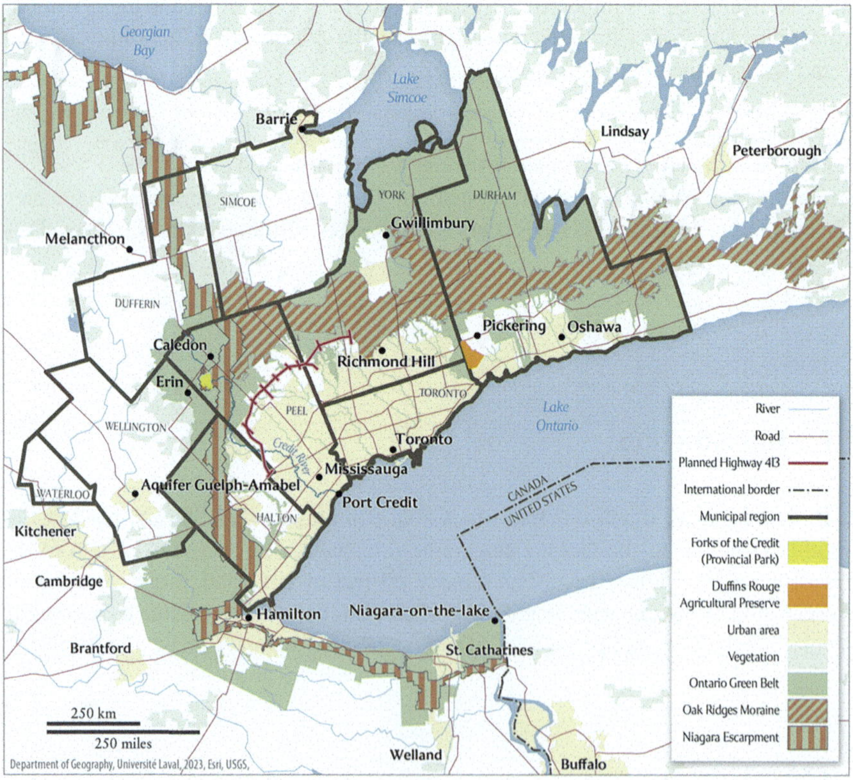

Fig. 3.1 The Ontario Greenbelt and the Oak Ridges Moraine in the Greater Toronto Area

Guelph-Amabel aquifer and serving as the headwater for five rivers supplying drinking water to one million Ontario residents and irrigating specialty crops, Melancthon township holds significant ecological importance (Suzuki, 2012; Government of Ontario, Environmental Registry website; Forster, 2017).

The proposed Melancthon Mega-quarry sought to convert 2316 acres (930 hectares) of prime farmland into the largest quarry project in Canadian history, reaching a depth equivalent to 20 storeys (Suzuki, 2012). The Highland Companies, backed by a $25-billion Boston hedge fund, initiated extensive land acquisitions, capitalizing on retiree farm owners after the 2008 global financial crisis when financial corporations sought the stability of land as a valuable asset (L'Italien, 2012). The company aimed to supply aggregates to meet the escalating demands of cement companies fueling the expansion of the Greater Toronto Area (Suzuki, 2012; Anders Sandberg et al., 2013).

Critical revelations emerged regarding the deceptive practices employed by the Highland Companies when acquiring farmlands. Agents assured landowners that their properties would retain their agricultural purpose, a claim disproven by subsequent evidence. In a CBC news documentary (Stiglic, 2011), many regretful farmers revealed that they would not have sold their lands had they known the true extractive intent of the purchases.

Journalists conducted inquiries that uncovered discrepancies in signed documents, a revelation that quickly gained traction in the media and alarmed the local population. The proposed limestone extraction project threatened to strip the prime potato-growing farmlands in Ontario of their topsoil, creating holes as deep as Niagara Falls. Despite the company's assurances of minimal impact on the watershed and promises of environmental restoration, skepticism arose. The ambitious plan involved storing the organic topsoil during extraction and returning it to the 200-feet deep hole once mining ceased, below the water table (Suzuki, 2012). Critics dismissed this plan as unrealistic, citing the challenges of restoring geological processes spanning 400 million years and reinstating the intricate natural ecosystems across multiple soil layers (Idem; Kuyek, 2019). Mining, viewed as a disruptive intervention in the land, leaves visible scars on landscapes and has profound implications for the affected communities.

Following the mobilization of engaged citizens in Melancthon township, bolstered by support from regional environmental advocacy groups focused on land and water protection, Toronto chefs and citizens from distant places like Toronto and Guelph rallied against the attempted land grab. Toronto chefs, aligning with the opposition movement, organized a demonstration, preparing soup in the fields earmarked for extraction, showcasing the local agricultural abundance (Elton, 2011). The opposition gained momentum as national organizations, including the David Suzuki Foundation, voiced their support, with numerous articles and letters to the editor in Canadian and American newspapers, such as Sarah Elton's piece in the Atlantic (2011), drawing attention to the contentious plan.

Amid growing dissent, the Ontario government, led by Premier Kathleen Wynne, intervened in September 2011, mandating a compulsory environmental assessment (EA) for the Highland Companies' proposal. Responding to perceived community disapproval, the company withdrew its mining license application on 21 November 2012, leading to the abandonment of the case (Halliday, 2012; Loney, 2012). This civic victory resonated nationwide, symbolizing Canadian civic society's resolute commitment to safeguarding agricultural lands against encroaching mining operations, with the once-pending Melancthon quarry becoming a compelling case study illustrating this determination.

3.3 Case Study No. 2: The GTA Greenbelt and Duffins Rouge Agricultural Preserve

With the resolution of the Melancthon Mega-quarry case, attention shifted back to everyday affairs at home and other contentious zones. However, on the southeastern horizon, whispers of a new highway project, Highway 413, began circulating. The proposed highway, set to be constructed north of Toronto, traverses through the Greenbelt and the Ontario Oak Ridges Moraine (see Fig. 3.1). The Oak Ridges Moraine, spanning over 160 km (99 miles) and shaped by glacier deposits from the last ice age, continues to attract interest from investors seeking agglomerates for construction. Cement industry companies are progressively acquiring substantial portions of the Toronto Greenbelt region, benefiting from previously secured mining rights.

Established in 2005, the Ontario Greenbelt stands as a permanently protected area, safeguarding environmentally sensitive land from urban development. As the largest protected Greenbelt globally, it surpasses counterparts in North America and Europe in terms of productivity (Coulter, 2022). The Greenbelt coincides in the north of Toronto with the Oak Ridges Moraine and joins the Niagara Escarpment to the west. It later curves southward and includes lands along Lake Ontario right till Niagara-on-the-Lake (the wine region).

When I returned to Ontario in 2022 and 2023, I saw road signs announcing the upcoming construction of Highway 413 northwest of Toronto, passing through Caledon. This new 59-km long (36.6 mi.) highway with 14 interchanges would devour around 400 acres of the Greenbelt, breaking Premier Doug Ford's 2018 promise to leave the Greenbelt untouched (Crandall, 2022). Despite a provincial environmental assessment since 2020, Caledon's town councilors in 2021 requested a federal environmental assessment and resident consultations (Town of Caledon website, 2021). The proposed highway has sparked concerns due to its potential impact on 29 threatened and endangered species, including notable ones like the Jefferson Salamander and the Eastern Musk Turtle, both classified as species of special concern. Additionally, worries extend to the significant number of river crossings and the potential harm to the Nashville Conservation Reserve, as highlighted by the environmental advocacy group Environmental Defence. Many local organizations and Caledon residents opposed the project. Environmental Defence argued that the underused 407 Highway could be an alternative, contradicting the

Ontario government's claim of a 30-min time saving for GTA residents, stating that Highway 413 wouldn't save any time.

Critics point out the outdated planning practices at a time when more sustainable, compact, and less fuel-dependent communities are recommended in the Golden Horseshoe Land-Use Plan (2021) (Crandall, 2022). Beyond consuming thousands of acres of prime farmlands with Class 1 soils, the proposed project threatens sensitive areas, including 220 wetlands and 85 streams in four watersheds (The Credit and Humber rivers and Etobicoke and Sixteen Miles creeks), as well as dozens of forests, woodland habitats, and cultural heritage sites of the Credit First Nation and Huron-Wendat archaeological sites (Crandall, 2022; Environmental Defence website).

It became evident to me that despite successfully resisting the initial project to convert potato lands into aggregates, the battlegrounds had multiplied across various sites. Another municipality, the town of Caledon, was now succumbing to corporate interests. This would not only divide the municipality with four exits built on its territory but also transform the town's landscape and disrupt the peaceful lifestyle of its residents. The Ontario Federation of Agriculture estimates that 175 acres of farmland are lost daily in the province due to development (Crandall, 2022). Suburban development in the Greater Toronto Area (GTA) has been a consistent trend for at least two decades, as documented in the Golden Horseshoe Growth Plan of 2006 (Gravel & Koné, 2017). The COVID-19 pandemic introduced a new pressure on peri-urban land, emphasizing the critical need for extensive storage space for companies like Amazon, leaving a noticeable impact on the suburban landscape.

3.4 The History of Land Protection Roll Back

In 2021, the Ontario government, under the Greenbelt, allowed the removal of protection from the Duffins Rouge Agricultural Preserve in Durham County, east of Toronto, using a special zoning order (MZO) to facilitate Amazon's warehouse construction in a wetland. Despite opposition from local conservationists and the Scugog Island First Nation, Amazon eventually chose an alternative site in the former Pickering Markets location (McConnell, 2022). While the land sale did not proceed, and the owners committed to its protection, the wetland was destroyed during land preparation, upsetting the Mississaugas of Scugog Island First Nation. They had expected the land to remain untouched, as it was designated for protection in 2005. The First Nation, along with local conservationist groups, discovered in November 2022 that the area had been cleared, possibly in response to the opposition from conservationists. This vulnerable land, originally protected from development, was not meant for construction according to the 2005 Act. The local First Nation expresses regret over the lack of respect for the land and highlights the existence of laws protecting wetlands, emphasizing that "the fight is not over yet" (McConnell, 2022).

In December 2022, the Ontario Government returned to charge and repealed the Duffins Rouge Agricultural Preserve Act of 2005 altogether by passing Bill 39, ironically entitled "The Better Municipal Governance Act".[1] On the Environmental Registry of Ontario under the title "Repeal Duffins Rouge Agricultural Preserve," the following can be read:

> The proposed *Duffins Rouge Agricultural Preserve Repeal Act, 2022* (DRAPRA) was introduced as Schedule 2 to Bill 39, *Better Municipal Governance Act, 2022* on November 16, 2022. If passed, the DRAPRA would:
>
> - end the deeming of easements or covenants that restrict the use of the lands to agricultural uses that apply to the Agricultural Preserve in the City of Pickering to be valid and allow for the amendment or release of these easements or covenants without the consent of the Minister under the *Conservation Land Act*.

In an alarming move, the repeal of the Duffins Rouge Agricultural Preserve Act in December 2022, facilitated by Bill 39, named "The Better Municipal Governance Act," was undertaken without any public consultation. This raises questions about the accuracy of the law's title, as it appears to be more about land management favoring developers' interests than governance improvement. Coupled with Bill 23, spearheaded by Ontario's Minister of Municipal Affairs, Steve Clark, these legislative actions have effectively opened significant portions of the previously protected Greenbelt (McConnell, 2022). The ambitious plan aims to construct 84,000 homes in the Regional Municipality of Durham, a region comprising nine municipalities, with Oshawa being the largest. The region already has an estimated population of 745,580, projected to double by 2032. This contradicts the intent of the revised Growth Plan, which emphasized the perpetual enhancement and protection of the Greenbelt's significant natural features, including the Oak Ridges Moraine and the Niagara Escarpment (Government of Ontario, 2020). Advocacy groups like "Stop Sprawl Durham" have joined forces with other environmental nongovernmental organizations (ENGOs) and the leader of the New Democratic Party (NDP) to challenge the government's decision to remove protection, holding them accountable for the potential environmental consequences (Jones, 2023).

The Duffins Rouge Agricultural Preserve, situated within the city of Pickering and covering 5200 acres of Crown land, plays a crucial role in local agriculture. Most farmers leased this land for cultivating grains intended for various purposes, such as animal feed, exportation, or ethanol production (Coulter, 2022). However, concerns have arisen regarding conventional and environmentally degrading agricultural practices that have reportedly led to contamination of underground and surface waters over the past five decades (Idem). Unlike the São Paulo Greenbelt in

[1] Duffins Rouge Agricultural Preserve's status in Ontario Environmental Registry is "Exception" since December 2022 (ERO website).

Fig. 3.2 (**a** and **b**) Whitevale road closure: A glimpse into the former Duffins Rouge Agricultural Preserve. (Credit: Gravel, 2023)

Fig. 3.3 (**a** and **b**) Rural tranquility in peril: Crop field on Whitevale Road in front of a condemned house. (Credit: Gravel, 2023)

Brazil, which is also a Biosphere Reserve, the Ontario government has not implemented stringent regulations to govern agricultural production within the Greenbelt, especially considering its critical role in supplying the Greater Toronto Area. The Duffins Rouge Agricultural Preserve is already the focus of residential developers, as evidenced by these pictures taken in June 2023 (Figs. 3.2a, b and 3.3a, b).

To enhance food production autonomy in the Greater Toronto Area, particularly highlighted by the COVID-19 pandemic, there is a need to value suburban food production within the Greenbelt and implement measures to protect and manage the remaining agricultural lands judiciously. The current cultivation of grains for export or ethanol production on prime Canadian lands may not be the most sustainable approach. A shift toward diversified production, steering away from monocropping, is essential to meet the needs of local food networks. This approach could benefit chefs, institutions, and city dwellers within a 100-mile radius (Idem; Elton, 2011). Coulter's work lays the ground for an environmentally sound proposal for Duffins Rouge agricultural lands, focusing on soil regenerative theories to breathe new life into these lands. However, realizing such a vision requires robust political will, with provincial stakeholders taking a bold stance to defend the last remaining prime agricultural lands and facilitate their recovery from decades of extractivist management practices.

3.5 Political Motives: Cement, Housing, and Alliances

The conduct of political affairs in Ontario, particularly under Premier Doug Ford (in power since 2018), has raised concerns among political analysts and journalists due to perceived discrepancies between official discourse and actions. The government's strong emphasis on economic development, encapsulated in Premier Ford's "open for business" agenda, has been associated with the influence of local land developers on land-use legislation (McConnell, 2022; Leffers, 2017). In response to the Ontario Housing Affordability Task Force's 2022 report highlighting a housing shortage, particularly in the Greater Toronto Area, Premier Ford prioritized the construction of 1.5 million homes, even if it meant removing existing environmental constraints (Government of Ontario, 2022; McConnell, 2022) (Figs. 3.4 and 3.5).

Fig. 3.4 Planned community north of Pickering, within the Greenbelt. (Credit: Gravel 2023)

Fig. 3.5 A street view in Richmond Hill, north of Toronto. (Credit: Gravel, 2023)
Description: Large houses fill most lots, leaving little room for greenery. All trees were removed during construction, with only a few replanted. The scene, with empty streets and a delivery truck, reflects a stark suburban post-pandemic reality

Despite the Ontario government's emphasis on expediting housing supply in 2022 to supposedly respond to growing immigrants' residential needs—a fact that has been proven wrong by Syed (2023) and McIntosh and Syed (2023)—, the Ontario Plan for Growth (2020) delineates principles aimed at cultivating "complete communities" that cater to diverse resident needs. These communities are encouraged to follow intensification and densification criteria, all while respecting the Greenbelt. Developers are urged to minimize land consumption through compact built forms, preserving agricultural lands, water resources, and natural areas. The plan also emphasizes the importance of protecting access to local and healthy food (Idem).

Development in the Peel region, notably in Mississauga, appeared to decelerate from 2016 to 2021. Conversely, there has been a notable uptick in growth along the North–South axis between Lake Simcoe and Richmond Hill, with rural communities like East Gwillimbury experiencing substantial population increases (Westoll, 2022). This burgeoning development trend might indicate an impending real estate surge north of Toronto in the forthcoming decade, potentially leading to what's colloquially termed "last harvests"—when farmers gather their final yields before selling their land (Rybczynski, 2007).

The loss of farmland to urbanization is certainly not new—with the proliferation of garden suburbs in the 1990s in the United States and Canada—but there's a shared feeling that the Anthropocene has put a heavy toll on farmlands and natural resources since the automobile revolution and what has escaped from its reach until now must be well cared after and managed with parsimony, like it were the last jewels of a crown. Natural habitats around large cities have been fragmented so much that Rachel Carson's prognosis of a silent spring (1962) is materializing itself. Cities and natural areas alike have become more silent as biodiversity declines at a dangerous rate. Over the past two decades, North America has witnessed a concerning decline of 30% in its backyard bird population, as noted by my ornithologist friend, Jean Denis Brisson, a trend corroborated by Mortillaro (2022). The 2022 World's Birds report further underscores this alarming trend, revealing that 49% of the world's bird species are experiencing decline and highlighting a looming biodiversity crisis (Idem). These statistics are profoundly unsettling and serve as a compelling impetus for action while there is still time. Emma Marris authored "Rambunctious Garden: Saving Nature in a Post-Wild World" in 2011 with the aim of catalyzing a grassroots movement to safeguard biodiversity in local environments, from backyards to roadside ditches. By piecing together these small fragments of habitat, we can create interconnected biodiversity corridors that support species living alongside us. Each report of bluebirds returning to suburban farms, thanks to citizens installing birdhouses, represents a small triumph worth celebrating! (See also range map and climate threat scenarios for bluebirds on the Audubon website.)

In my quest to understand the industrial activities within the Ontario Greenbelt, I turned to watershed conservationist groups for insights. To my surprise, I found these groups grappling with the pervasive influence of economic forces, primarily driven by two factors: (1) the existence of preexisting extraction rights granted before the enactment of the 2001 Oak Ridges Moraine Act and the 2005 Greenbelt Act; and (2) a significant reliance on funding from the provincial government, rendering them dependent on it for project continuity (e.g., Fig. 3.6). As a result, some conservationists find themselves constrained to employ a fragmented project management approach, which hampers their ability to comprehend the broader regional perspective and critically assess the detrimental impacts of land-use changes. Therefore, authentic opposition must stem from citizen-led and independent advocacy groups to ensure a more comprehensive and effective response. The management of the Toronto Greenbelt involves a strategy of land exchanges, where some lands are earmarked for exploitation, while others are strategically negotiated and fully safeguarded, even if located outside the official perimeter (Leffers, 2017).

When discussing particularly sensitive areas such as headwaters, I was taken aback to learn from two conservationists that they "ignored" a conspicuous limestone extraction site adjacent to the entrance of Forks of the Credit Provincial Park (see Fig. 3.1). Such revelations prompted doubts regarding the sincerity of their responses. It begs the question: Does environmental protection in Ontario currently operate in such a compartmentalized and fragmented manner, driven by project management and short-term goals, that it obstructs a holistic understanding of

Fig. 3.6 Multiscale partnership announcement in the Credit River Valley. (Credit: Gravel, 2022)

the broader situation? Situated within the Niagara Escarpment and the Ontario Greenbelt, this park exemplifies the complexities and challenges inherent in preserving such crucial natural landscapes, as illustrated in Figs. 3.7 and 3.8 below.

The unmistakable presence of numerous aggregates-laden trucks on the nearby road, accompanied by the dust clouds they generate and the visibly affected fields, becomes impossible to ignore while driving toward the provincial park. Considering the known impacts of such industry on water quality and fish habitats, the coexistence of these activities in the headwaters of the Credit River raises questions, especially considering that this river is recognized for being one of the best protected in the Greater Toronto Area and is often considered a model for watershed governance. The Oak Ridges Moraine plays a crucial role in filtering water through its multiple-layered glacial stone deposits and storing underground water. As this water emerges on the surface, it becomes the source of streams and rivers. It is evident that these areas are vital for aquifer recharge and maintaining water quality in the streams below. Any form of mechanical deforestation and extraction involving heavy equipment is bound to alter the nature of adjacent lands, stripping them of vegetation and compacting the soil. This, in turn, disrupts the hydrological flux, influencing how water travels and is stored in the ground, eventually reaching rivers and streams. The presence of trees and natural vegetation along riverbanks helps mitigate water flow velocity, promotes infiltration into the ground, and prevents surface runoff that can cause erosion and transport sediments into the streams (McIntosh, 2021).

Fig. 3.7 Interpretative Sign at Forks of the Credit Provincial Park. (Credit: Gravel, 2022)

Fig. 3.8 Aggregate truck crossing neighboring the Forks of the Credit Provincial Park, Caledon Village. (Credit: Gravel, 2022)

Despite preexisting mining rights, proactive initiatives led by the Niagara Escarpment Biosphere Reserve, the Ontario Greenbelt, the provincial government, and the town of Caledon could lead to the acquisition of lands situated in sensitive areas. Following the principles of the latest ecological research, these efforts could include restoration measures if any damage has been incurred in the headwaters (Coulter, 2022). A notable example is the Save the Redwoods League in California, which successfully advocated for the purchase of 400 acres of forested lands along the Russian River to establish an ecological corridor and safeguard pristine riparian forest lands on the brink of logging (further details provided at the end of this chapter). The positive outcome of this citizen-led initiative could serve as inspiration for similar groups globally.

Ideally, action should be taken before any vegetation removal occurs, emphasizing the importance of developing a municipal regulatory framework that aligns with the preferences of the local community. Kuyek (2019), who has supported citizens and First Nations in Canada in defending their rights against the mining industry, argues that Ontario's Mining Act is outdated and requires modernization to better address the concerns of resident populations, including First Nations. Municipal mining laws should also be reconsidered to empower city councilors in preserving the quality of life for the constituents they represent.

In addition to the existing sand and gravel pits, many of which are owned by Holcim, a Swiss company and the largest landholder in Caledon (Ross, 2022), there is an ongoing proposal for a new quarry in the town. Concerned residents have united under the banner of the Forks of the Credit Preservation Group (FCPG) to resist the planned quarry, which is backed by the Brazilian multinational company Votorantim Cementos. The FCPG website states:

> In 2019, we were shocked when they learned of the plan by a Brazilian conglomerate, Votorantim Cimentos/CBM St Marys, to locate a mega blasting quarry on 800 acres of prime agricultural land adjacent to the Credit River. Differing from gravel pits, this was a plan to use explosives to blast 80 feet below the water table and release the rising water into the river. Disbelief was soon replaced by resolve when David Sylvester and his neighbors created the Forks of the Credit Preservation Group to protect the place we all deeply love. We sounded the alarm and were joined by hundreds of others with the same concerns.

Residents are expressing concerns over potential threats to the community posed by the proposed quarry. These concerns include issues such as fly rocks resulting from blasting at an unregulated distance from human activities (as the Caledon municipality lacks legislation on mining activity near human settlements), increased noise, truck traffic, and the risk of drilling below the aquifer level. The proposed quarry site is situated less than 1 km (0.6 mi) from the Forks of the Credit River, raising fears of possible contamination through leaching and sedimentation that could adversely affect the river's ecosystem, particularly native species like the speckled trout (brook trout) and the redside dace, the latter being on Ontario's endangered species list since 2009 (FCPG website; McIntosh, 2021).

3.6 Discussion

Comparative research by the Forks of the Credit Preservation Group (FCPG) highlights Caledon's weak legislative framework on mining compared to other Ontarian cities. As a result, they, like many other groups impacted by the extraction industry, called for a review and modernization of the old Ontario Mining Law (Idem; Kuyek, 2019). This prompts the question of how effective the dual protection provided by the Niagara Escarpment Commission and the Ontario Greenbelt is in preserving the integrity of this ecologically sensitive territory.

Joan Kuyek, the author of "Unearthing Justice: How to Protect Your Community from the Mining Industry," offers valuable advice for communities opposing mining projects: develop an alternative vision. Kuyek argues that mining gains support because people believe it's the sole means of bringing cash and jobs to the community (Idem: 236). Successful community resistance against mining often hinges on demonstrating two key points: (1) there's already an existing economic plan benefiting the affected area significantly, and (2) the proposed mine would irreparably harm that plan (Ibid.). Protecting farmlands from corporate interests can involve establishing farmland trusts, similar to those in the United States or the agroecological social utility trust in Quebec. These trusts reserve lands for farmers to lease under their conditions, ensuring the ongoing preservation of agricultural production (Mesly, 2022; L'Italien, 2012).

Aggregate extraction stands out as one of Ontario's most robust economic sectors. While mining proponents tout local economic benefits and contributions to the GDP, they often overlook external costs. In the case of transnationals, these costs are deferred to present and future generations (Kuyek, 2019). Communities bear a significant share of these costs, experiencing impacts on their lifestyle (noise, dust, traffic, contamination, property expropriations) and environmental degradation (air, water, soil quality loss, and a diminished sense of peace). Taxpayers also contribute through subsidies and power rebates that benefit the companies. Kuyek stresses that the mining industry operates within an "economy of loss" in Canada, where profits are derived from environmental and societal degradation (Idem: 2).

Numerous locally owned businesses have engaged in aggregate extraction within the Oak Ridges Moraine region for decades, contributing both employment opportunities and revenue to the provincial GDP (Ross, 2022). However, the industry has also attracted international investors, such as the Canadian branch of the Brazilian conglomerate Votorantim Cemento, which acquires farmland and implements a "scalping technique."

This process involves stripping the top 30 cm (12 inches) of soil to access the desired rock material, leaving behind a landscape reminiscent of Swiss cheese (Mesly, 2022). The juxtaposition of active gravel pits and farmland in a confined space creates challenges for agricultural activities and disrupts the coexistence of these land uses. In the town of Erin, Ontario, near Caledon, a series of images taken in August 2022 captures this transformation, with the Ontario Federation of

Agriculture sign still visible on the fence, showcasing the corn variety that once thrived on the now-altered land (Fig. 3.9a–c).

Until September 2023, it seemed unlikely that these examples would demonstrate the strength of the Canadian civil society, unlike the success in Melancthon. The Ontario government, led by Premier Doug Ford, appeared more focused on economic development and aligned with land developers. Doug Ford even mentioned in a video that the idea of opening up the Greenbelt was not his, but a developer's, although he expressed a desire for it to be his own idea. However, things changed after the release of a report in September 2023 by the Ontario general auditor and integrity commissioner, which exposed collusion and the illegal removal of Greenbelt protections. Following this revelation, Prime Minister Doug Ford publicly apologized for his actions, and the Ontario Housing Minister resigned (McIntosh & Syed, 2023; Jones, 2023). This shift was influenced by the work of local and provincial environmental groups (ENGOS) and investigative reporters, such as those from The Narwhal, who uncovered well-connected developers buying lands in the Duffins Rouge Agricultural Reserve, anticipating the removal of legal protections. While opponents of Highway 413 hope this lesson will also stop the project, concerns remain. Despite the Ontario Premier's promise to return lands to the Greenbelt and protect it, the fate of the GTA western transit project is still uncertain, pending conclusions from the Environmental Protection Agency (EPA) (Idem).

Regarding the Caledon quarry, there is still the potential for a civic triumph. It seems more feasible to rally communities and environmentalists against a corporate initiative, especially if it stems from a multinational or foreign corporation, than to garner support for contesting a government decision, particularly in the absence of cooperation from authorities. An effective preventive measure is to ensure the election of candidates who prioritize the public's interests for the upcoming election.

Opponents of Highway 413 frequently advocate for the protection of local food production and enhancing food security. Unlike some other Canadian provinces such as Quebec and British Columbia, Ontario lacks a commission dedicated to protecting farmland. In the past 30 years, the Greater Toronto Area (GTA) and the Niagara Peninsula have seen a lot of changes. Valuable farmlands have been replaced by big buildings or fancy housing projects, with names recalling the lost natural features. The rise of vineyards, especially in the "Wine country," has made it tough

Fig. 3.9 **(a–c)** Visual contrast: Two adjacent farmlands—one inhabited and productive, the other stripped for aggregates. (Credit: Gravel, 2022)

for small fruit growers (see Sect. 3.7). This has led to the decline of orchards in the Greenbelt, affecting people across Canada. Fruit trees unique to Central Canada are also at risk due to urban expansion. Many argue that protecting farmlands is a national concern and should take precedence over urban development (Mesly, 2022). Using good farmland for buildings and parking lots is seen as a poor use of land. Ontario's land-use planning needs a longer-term and sustainable approach, including stopping the reclassification of agricultural or forested lands through special decrees.

The planned housing developments at the outskirts of the GTA and the proposed construction of Highway 413 are expected to increase the need for building materials in environmentally important areas like the Oak Ridges Moraine and the Greenbelt. This rising demand may lead to a relaxation of existing environmental protections in these regions (Jones, 2023). Currently, 39% of active material extraction sites are within the Greenbelt, and the town of Caledon is the closest source for materials for the new highway. However, this will bring more traffic to Caledon, as it is set to host 8 out of 14 planned highway exits and an estimated population of 300,000 by 2050. This significant change contrasts with Caledon's current population of 81,903 in 2022, with an annual population growth rate of +3.1% in recent years (Crandall, 2022). The potential impact on Caledon depends on whether the concerns raised by the Save the Oak Ridges Moraine Coalition,[2] a group of 25 citizens' groups active since 1989, are considered. These changes would contradict the efforts by Caledon's leaders to aim for a net-zero carbon target by 2050 and implement a climate action plan focusing on a resilient food and agriculture sector, along with the protection of natural lands (Crandall, 2022; Coulter, 2022).

Mitigating the Ecological Footprint of Winemaking in the Niagara Peninsula

Situated in the southern part of the Ontario Greenbelt, the Niagara Peninsula offers an exceptional climate for vine cultivation which is poised to gain momentum with climate change and the warming of nordic regions (Cukierman et al., 2021) (see Niagara-on-the-lake on Fig. 3.1 which is at the heart of Ontario's Wine Country). However, the expansion of vineyards in the Niagara Peninsula has exerted a significant ecological strain in the past on Lake Ontario, one of the Great Lakes. This strain was primarily due to the extensive footprint resulting from monoculture farming practices—land clearing, chemical treatments—and the competition with food production for

[2] STORM, the Save the Oak Ridges Moraine Coalition, operates a Moraine Interactive Mapping Project aimed at providing tools and information to concerned citizens about development activities within the Moraine (see its website https://www.stormcoalition.com/copy-of-morain-watch).

land use. In response, a growing awareness among vineyard owners led to the creation of the Ontario Sustainable Winegrowing Certification in 2007. Despite the increasing number of vineyards in the region,—now exceeding 50—the goal is for all winegrowers to achieve certification by 2030 (Sustainable Winegrowing Ontario Certified website).

Among them, Southbrook Vineyards stands out as a remarkable leader since 2005, boasting certifications for sustainable vineyards and winemaking, organic farming (Ecocert Canada), and green building design (LEED) and showcasing a dedication to environmentally responsible practices (Southbrook Vineyards website). Beyond pioneering Canada's first orange wine from grape peels, Southbrook Vineyards has embraced sustainable practices encompassing efficient water usage, organic farming without chemical pesticides or synthetic fertilizers, and the integration of sheep and hens for natural fertilization and pest control. Additionally, it has invested in solar energy utilization. Over the years, Southbrook has continued to implement various eco-friendly techniques, including wetland water treatment, agroforestry[3] with fruit trees and bushes, and biodynamic winemaking to uphold the vitality of their wines, alongside solar power generation.

By implementing these strategies and fostering a culture of sustainability within the wine industry, it is possible to diminish the ecological footprint of winemaking in the Niagara Peninsula while still producing high-quality wines. However, implementing more stringent provincial and municipal legislation could help protect the remaining non-vineyard agricultural lands and green spaces, which are crucial for biodiversity conservation in the area.

3.7 Urban Growth Regulation: Brief Exploration of Alternative Models in America

In 1972, Petaluma pioneered growth control measures by becoming the first city in the United States to restrict annual building permits. The city also implemented an urban-growth boundary. Over time, these innovative planning strategies, collectively recognized as "smart growth," gained widespread adoption by numerous cities across the country (Adrianson, 2009).

The Petaluma case study in California during the 1970s serves as a noteworthy illustration of sustainable urban planning strategies. In response to the imperative of curbing unchecked urban growth and preserving the distinct identity of the town,

[3] Agroforestry is the science that studies the combination of trees and crops in functional systems designed to promote ecological sustainability in riverbanks, farmlands, and agroecosystems.

Petaluma embraced a policy of zero growth. This deliberate decision to limit population expansion was accompanied by the establishment of a greenbelt around Petaluma and other cities within Sonoma County. The greenbelt not only served as a physical barrier to urban sprawl but also functioned as a proactive measure to safeguard the unique character and cultural heritage of these towns. This strategic approach stands in stark contrast to the urban sprawl evident in the Bay Area of San Francisco, highlighting the efficacy of thoughtful and proactive planning in maintaining the integrity and identity of communities amid the dynamic forces of urban development.

Sonoma County has held onto its agricultural heritage by design, unlike other parts of the Bay Area that experienced urban sprawl. At the end of the 1990s, voters in Sonoma County passed an ordinance to prevent most types of development in the spaces between cities, ensuring the preservation of the region's rural character. This deliberate action has effectively prevented excessive urbanization and supported the longevity of Sonoma County's agricultural legacy.

Addressing calls from developers eager to build houses in the greenbelt, Teri Shore, the North Bay director of the Greenbelt Alliance, emphasized the protective nature of the measure: "This measure protects open space and farmland. It prevents the conversion to shopping malls and housing tracts without the vote of the residents of Sonoma County. It holds back sprawl for the next generation" (Brown, 2016). In response to considerations of the land between cities as a potential solution for affordable housing, Shore asserted that there is sufficient space within existing city limits for such projects, citing ongoing or approved initiatives in Petaluma. Contrary to the notion that greenbelts exacerbate the housing crisis, Shore asserted: "Greenbelts did not cause the housing crisis, and building in greenbelts won't solve the housing issue" (Idem). "We've avoided the mistakes made in other Bay Area counties that go from one city to another without any identity," Shore said.

An additional example involves the Save the Redwoods League in California, which purchased approximately 400 acres of forested terrain along the Russian River, located in the historical territory of the Pomo people, in October 2023. This acquisition aims to preemptively shield these lands from impending logging activities (Idem; Save the Redwoods League website; Guerneville Forest Coalition website). Revised: The former owner had outlined plans to harvest 200 acres of this woodland by the coming year, despite opposition from local community groups. Within this expanse, nestled amid Douglas firs and coast redwoods, stands the Clar Tree, a remarkable specimen with a 16-foot diameter and an impressive height of 278 feet. Estimated to be around 2000 years old, the Clar Tree is recognized as the tallest tree in Sonoma County and is cherished by the local community of Guerneville. The conservation effort aimed not only to protect the area's rich biodiversity but also to preserve the cultural and ecological legacy embodied by this ancient tree (Callahan, 2023). After a 3-year fight, the Guerneville Forest Coalition, a group of engaged citizens gathered under the Save the Redwoods League, rapidly raised the necessary funds to save this patch of redwoods and the one-mile Russian River coastal area. This effort is essential to protect the water of the Russian River and its population of anadromous fish (Guerneville Forest Coalition website; Save the Redwoods League, 2023 video).

The acquisition of the Russian River Redwoods, valued at $6.5 million, marks a crucial step in establishing an ecological corridor (Idem). This property includes a half-mile stretch along Mays Canyon Creek, a habitat for steelhead trout, and is adjacent to the Bohemian Grove, owned by the exclusive Bohemian Club. Combined with existing lands, it forms a contiguous 4000-acre wildlife corridor. On its website, under the title "Rewriting the story of Russian River Redwoods" (2024), it states, "The League intends to transfer the land to Sonoma County, in partnership with the Sonoma County Agricultural Preservation and Open Space District, for long-term stewardship, forest restoration, and potential access."

3.8 Conclusion

Environmental citizenship is demonstrated through active civic engagement and originates from a profound connection to a particular location, representing a dedication to maintaining the integrity of one's environment. It fosters a sense of civic identity across various geographic levels and motivates involved citizens to take action. Recent challenges in Ontario, Canada, such as proposals to establish a quarry in valuable farmlands for extraction and efforts to remove protection from preserved areas, faced strong opposition from knowledgeable and organized citizens.

The Rouge National Urban Park, located along the Rouge River and home to the Toronto Zoo,[4] serves as a critical biological corridor connecting to the Duffins Rouge Agricultural Preserve. The planned removal of the preserve's protection, coupled with land swaps for development within the Greenbelt and the Oak Ridges Moraine, has raised concerns about the impacts on endangered species and the well-being of nearby First Nations communities. These concerns have prompted the Federal government's Environment Ministry to contemplate intervention.

Environmental NGOs, together with empowered citizens in Pickering and Toronto, successfully thwarted the threat to these preserved lands in September 2023, using well-documented analyses from independent local journalists. In an uncommon turn of events, Premier Doug Ford was pressured by the Ontario Auditor General's office to reverse his previous decisions and continue protecting the lands designated for conservation.

While celebrating this citizens' victory, it remains prudent to explore and implement alternative strategies to current land conservation methods. Drawing inspiration from successful initiatives, such as the ongoing efforts to secure the Russian River Redwoods, is advisable given the persisting vulnerability of protected lands

[4]The Toronto Zoological Garden was honored with the 2023 Canada's Greenest Employer award and has been implementing a sustainability plan since 2007 (The Toronto Zoo website).

in the GTA. Valuable lessons should be learned to devise more robust and enduring protection mechanisms while wisely utilizing their agricultural potential. Embracing approaches such as agroforestry, agroecology, conservation land trusts, and payment for ecological services may offer constructive solutions to navigate challenges posed by influential figures like Premier Doug Ford and well-connected developers in the future.

References

Adrianson, A. (2009). A pioneer in smart growth becomes a pioneer in no growth. *The Daily Signal*. Retrieved at: https://www.dailysignal.com/2009/05/07/a-pioneer-in-smart-growth-becomes-a-pioneer-in-no-growth/. Accessed 18 Jan 2024.

Anders Sandberg, L., Wekerle, G. R., & Gilbert, L. (2013). *The Oak Ridges Moraine Battles. Development, Sprawl, and nature conservation in the Toronto Region*. University of Toronto Press.

Brown, M. (2016). County Greenbelt on November Ballot. *Petaluma Argus Courier*. Retrieved at: https://www.petaluma360.com/article/news/county-greenbelt-on-november-ballot/. Accessed 18 Jan 2024.

Callahan, M. (2023). Save the Redwoods League to buy nearly 400 acres of redwood forest along lower Russian River. *Petaluma Argus Courier*. Retrieved at: https://www.petaluma360.com/article/news/save-the-redwoods-league-to-buy-nearly-400-acres-of-redwood-forest-along-lo/?artslide=0. Accessed 18 Jan 2024.

Carson, R. (1962). *Silent spring*. Houghton Mifflin.

Coulter, S. A. (2022). *In defense of soil: An Ontario Greenbelt Regeneration strategy*. Master's thesis, University of Waterloo Space. Retrieved at: https://uwspace.uwaterloo.ca/handle/10012/18067. Accessed 18 Jan 2024.

Crandall, D. (2022). Why Highway 413 is a bad idea. *In the Hills*. Retrieved at: https://www.inthehills.ca/2022/03/why-highway-413-is-a-bad-idea/. Accessed 18 Jan 2024.

Cukierman, J., Quénol, H., & Bouffard, M. (2021). *Quel vin pour demain? Le vin face aux défis climatiques. Paris, Dunod.*

Elton, S. (2011). Foodstock: Canadian foodies and Chefs Fight Mega Quarry in Ontario. *The Atlantic Monthly*. Retrieved at: https://www.theatlantic.com/health/archive/2011/10/foodstock-canadian-foodies-and-chefs-fight-mega-quarry-in-ontario/247339/. Accessed 18 Jan 2024.

Forster, P. (2017). Escarpment use and population growth: Environmental and social sustainability policies within the Niagara Escarpment. *Sustainable Heritage Case Studies*. Retrieved at: https://sustainableheritagecasestudies.ca/2017/12/12/escarpment-use-and-population-growth/. Accessed 18 Jan 2024.

Government of Ontario. (2020). *A place to grow. Plan for the Greater Golden Horseshoe*. Office Consolidation 2020. Retrieved at: https://www.ontario.ca/document/place-grow-growth-plan-greater-golden-horseshoe. Accessed 18 Jan 2024.

Government of Ontario. (2022). *Report of the Ontario Housing Affordability Task Force*. Retrieved at: https://files.ontario.ca/mmah-housing-affordability-task-force-report-en-2022-02-07-v2.pdf. Accessed 18 Jan 2024.

Gravel, N., et Koné A. (2017). The Guelph water connection: The contribution of actor-network theory (ANT) to the study of water management in Guelph, Ontario. *Les Cahiers de géographie du Québec*. Special issue on water governance in Canada and Brazil. Ed. L. Guay and A. Britto. 61(174), 489–512. Retrieved at: https://www.erudit.org/fr/revues/cgq/2017-v61-n174-cgq04102/1053663ar.pdf. Accessed 18 Jan 2024.

Halliday, C. (2012). Update: Melancthon quarry plan is no more. *Orangeville*. Retrieved at: https://www.orangeville.com/news-story/1482555-melancthon-quarry-plan-is-no-more/. Accessed 18 Jan 2024.

Jones, R. P. (2023). 1,200 homes on prime farmland: What's known about the plan to develop the Greenbelt in Pickering, Ont. *CBC News*. Retrieved at: https://www.cbc.ca/news/canada/toronto/duffins-rouge-agricultural-preserve-1.6937144. Accessed 18 Jan 2024.

Kuyek, J. (2019). *Unearthing Justice. How to protect your community from the mining industry*. Between the Lines.

L'Italien, F. (2012). *L'accaparement des terres et les dispositifs d'intervention sur le foncier agricole : les enjeux pour l'agriculture québécoise*. Institut de recherche en économie contemporaine. IRÉC Research Report. Retrieved at: https://numerique.banq.qc.ca/patrimoine/details/52327/2772038. Accessed 18 Jan 2024.

Leffers, D. (2017). Developers, the State, and the politics of private property rights. *Journal of Law and Social Policy, 27*, 165–194.

Loney, H. (2012). Highland companies withdraws mega-quarry proposal for north of Toronto. *Global News*. Retrieved at: https://globalnews.ca/news/310987/highland-companies-withdraws-mega-quarry-proposal-for-north-of-toronto/. Accessed 18 Jan 2024.

Marris, E. (2011). *Rambunctious garden: Saving nature in a post-wild world*. Bloomsbury.

McConnell, L. (2022). "We Were Shocked": Mississaugas of Scugog "appalled" by wetland destruction. *InSauga*. Retrieved at: https://www.insauga.com/we-were-shocked-mississaugas-of-scugog-appalled-by-wetland-destruction-in-pickering/. Accessed 18 Jan 2024.

McIntosh, E. (2021). Everything you need to know about the endangered species, waterways and farmland in southern Ontario's Greenbelt. *The Narwhal*. Retrieved at: https://thenarwhal.ca/ontario-greenbelt-explainer/. Accessed 18 Jan 2024.

McIntosh, E., & Syed, F. (2023). Doug Ford reverses Greenbelt decision. *The Narwhal*. Retrieved at: https://thenarwhal.ca/ford-reverses-greenbelt-decision/. Accessed 18 Jan 2024.

Mesly, N. (2022). *Terres d'asphalte. Notre agriculture sous haute pression*. Éditions MultiMondes.

Mortillaro, N. (2022). Nearly half of the world's birds are on the decline, which experts say is a serious threat to ecosystems. *CBC News*. Retrieved at: https://www.cbc.ca/news/science/birds-declines-1.6610220. Accessed 13 Jan 2024.

Ross, N. (2022). Pit by Pit. *In the Hills*. Retrieved at: https://www.inthehills.ca/2022/09/pit-by-pit/. Accessed 18 Jan 2024.

Rybczynski, W. (2007). *Last harvest. How a Cornfield became New Daleville*. Simon & Schuster Inc.

Stiglic, J. (2011). Quarry on Ont. Farmland was the plan, firm says. *CBC News*. Retrieved at: https://www.cbc.ca/news/canada/quarry-on-ont-farmland-was-the-plan-firm-says-1.1077670. Accessed 18 Jan 2024.

Suzuki, D. (2012). Melancthon mega-quarry: It's limestone vs. potatoes. *The Toronto Star*. Retrieved at: https://www.thestar.com/opinion/melancthon-mega-quarry-it-s-limestone-vs-potatoes/article_08dc6622-9b45-5836-8013-0a25dc1232ad.html. Accessed 18 Jan 2024.

Syed, F. (2023). Immigrants are not the reason Ontario's Greenbelt is being developed. *The Narwhal*. Retrieved at: https://thenarwhal.ca/ontario-greenbelt-immigration/. Accessed 18 Jan 2024.

Westoll, N. (2022). Census 2021: How population levels in the Greater Toronto Area changed in 5 years. *City News Everywhere*. Retrieved at: https://toronto.citynews.ca/2022/02/10/census-2021-canada-toronto-population/. Accessed 24 May 2024.

Websites

Aububon website. Eastern bluebird. Accessed August 31, 2024. https://www.audubon.org/field-guide/bird/easternbluebird

City Population- Regional municipality of Durham. https://www.citypopulation.de/en/canada/ontario/admin/3518__durham/. Accessed 18 Jan 2024.

City Population- Town of Caledon. https://www.citypopulation.de/en/canada/ontario/admin/peel/3521024__caledon/. Accessed 18 Jan 2024.

Environmental Defence. 10 endangered species that will suffer if Highway 413 is built. May 20, 2022. https://environmentaldefence.ca/2022/05/20/10-endangered-species-that-will-suffer-if-highway-413-is-built/. Accessed 18 Jan 2024.

Forks of the Credit Preservation Group (FCPG). *About us*. https://www.fcpreservation.ca/about_us. Accessed 18 Jan 2024.

Government of Ontario environmental Registry – Niagara Escarpment. https://ero.ontario.ca/index.php/fr/comment/3759. Accessed 18 Jan 2024.

Guerneville Forest Coalition – The Clar Tree. https://www.guernevilleforestcoalition.org/. Accessed 24 May 2024.

Ontario Environmental Registry (ERO) Duffins Rouge Agricultural Preserve. https://ero.ontario.ca/notice/019-6304. Accessed 18 Jan 2024.

Save the Redwoods League. (2023). *Video: Russian River Redwoods*. https://www.youtube.com/watch?v=W82VYALMmKM&ab_channel=SavetheRedwoodsLeague. Accessed 24 May 2024.

Save the Redwoods League. (2024). *Rewriting the story of Russian River Redwoods*. https://www.savetheredwoods.org/project/russian-river-redwoods/rewriting-the-story-of-russian-river-redwoods/. Accessed 24 May 2024.

Southbrook Vineyards. https://www.southbrook.com/About/Green-Map. Accessed 24 May 2024.

STORM, Save the Oak Ridges Moraine Coalition. https://www.stormcoalition.com/copy-of-morain-watch. Accessed 24 May 2024.

Sustainable Winegrowing Ontario Certified website. Accessed August 31, 2024. https://sustainablewinegrowingontario.ca

Toronto Zoological Garden. Going Green. *2023 Canada's Greenest Employers*. https://www.torontozoo.com/green. Accessed 19 Jan 2024.

Town of Caledon – GTA West Transportation Corridor. https://www.caledon.ca/en/government/gta-west-transportation-corridor.aspx, Accessed 18 Jan 2024.

Chapter 4
Striving for Harmony: The Reventazón Model Forest of Costa Rica

Abstract In Costa Rica, the traditional centralized approach to natural resource management has been challenged by the establishment of the Reventazón Model Forest (RMF) in Cartago Province. This new model of forest landscape governance empowers local communities and stakeholders to actively participate in collaborative territorial management. Operating within the International Model Forest Network (IMFN), the RMF facilitates collaboration among government, private sector, and local populations, representing a significant step toward community autonomy. Importantly, the RMF's inclusive efforts extend to Jameikari communities, acknowledging their expertise and involving them in activities such as cultural tourism guided by indigenous youth. The IMFN's horizontal structure offers an alternative approach to governing forest landscapes and watersheds in Costa Rica, addressing limitations of conventional debt-for-nature agreements that have shaped the country's national parks map. Alongside legislative measures, the RMF has contributed to Costa Rica's significant biodiversity recovery.

Keywords Biodiversity · Cartago Province · Costa Rica · Cultural tourism · Decentralization · Environmental management · Landscape management · Indigenous communities · International Model Forest Network (IMFN) · Indigenous communities · Local communities · Participatory governance · Comanagement · Inclusion of marginalized groups · Debt-for-nature swap · Reventazón Model Forest (RMF)

4.1 Introduction

Environmental management is undergoing a significant transformation as the principles of governance are starting to be incorporated into landscape management in various parts of the planet, with the decentralization of state power being its most visible manifestation (Blaser et al., 2005). While forest management in Latin America, as well as in other regions of the world, has long been the prerogative of central governments since the colonial era, it now increasingly falls under

N. Gravel, *Tending Nature*, https://doi.org/10.1007/978-3-030-47062-3_4

subnational governments. These subnational governments play an increasingly important role in the creation and implementation of forest and environmental conservation policies. Local populations, along with international organizations, nongovernmental entities, and other civil society organizations, are advocating for a say in forest management, considering it a common good and a global heritage (Capistrano & Pierce Colfer, 2005; Ros-Tonen et al., 2007).

4.2 Among the Worlds' Champions in Forest Protection

According to a 2023 OECD report, Costa Rica is well-positioned among the world's countries to achieve Sustainable Development Goals (SDGs) and transition to a net-zero economy in the coming decades. Its "Sustainable Landscape Initiative 2022–2030" aims to reduce emissions from agriculture, forestry, and other land uses, aligning with its commitments under the 2021 Glasgow Declaration on Forests and Land Use. Despite this progress, challenges remain: urban air quality is degraded in the Greater Metropolitan Area of San José, where 73% of the population resides, and there is a need for improved public transportation to reduce emissions (OECD, 2023). Additionally, road safety must be enhanced to encourage cycling and walking commutes, while water pollution and declining water quality in rivers pose significant environmental concerns in Costa Rica (OECD, 2023).

Forest coverage in Costa Rica has experienced remarkable growth over the last 20 years, increasing from 21% in 1987 to an impressive 59% today, earning the country the distinction of being the first tropical nation to halt deforestation (OECD, 2023). Approximately half of these forested areas benefit from some form of protection, ranging from biological corridors to national parks, although the level of protection varies widely. This achievement can be primarily attributed to a combination of policies introduced since 1995 aimed at environmental conservation and enhanced forest management, including the enactment of the Forestry Law in 1996, coupled with the collapse of the beef market in the 1980s (Idem). Several innovative measures have been implemented to encourage sustainable use, restoration, and conservation initiatives for forests, including Payments for ecosystem services (PES)[1] (Le Coq et al., 2012; Robalino & Pfaff, 2013). Despite a reduction in deforestation during the 1980s and 1990s, the forested area hit its nadir in 1987 at 21% (Fonafifo, 2005, cited in Le Coq et al., 2012). The enactment of the Forestry Law in 1996, accompanied by various incentives for forest conservation, including the initiation of PES in 1997, and the thriving ecotourism sector, coupled with reduced investments in cattle ranching and agriculture, collectively created a conducive environment for reversing the trends of deforestation (Ibid).

[1] Payments for environmental services (PES) involve individuals or groups receiving compensation from a funding partner (such as the World Bank) for activities that benefit the environment, such as conservation efforts or sustainable resource management.

Despite progress in restoration and reforestation initiatives, forest fragmentation remains a significant concern. While dry forests have shown signs of recovery, humid and cloud forest ecosystems require more time to regenerate following clearance, and second-growth forests often lack the biodiversity of their original counterparts. Additionally, forest habitats face increasing threats from wildfires and climate change, necessitating careful management practices to mitigate these risks (Idem).

Furthermore, challenges persist due to factors such as population growth, urbanization, and tourism, which place additional strain on the environment and contribute to issues like water pollution and waste generation. Addressing these challenges will require concerted efforts and innovative strategies to ensure the long-term sustainability of Costa Rica's forests and ecosystems.

Costa Rica's legal framework prioritizes environmental protection, and the government has demonstrated innovation in safeguarding its coastline[2] and oceans. In 2019, Costa Rica made history by enacting legislation to protect its coral ecosystems, setting a precedent for other nations (AIDA Press, 2019). While these laws may sometimes be imperfect and challenging to enforce, municipalities play a crucial role in ensuring compliance (Cabrera, 2008). Overall, Costa Rica's efforts have been effective in preserving its natural heritage and regulating tourism development (Montero, 2022). In a 2022 Gallup survey, 67% of citizens expressed a high level of satisfaction with the government's efforts to protect the environment. Additionally, support for environmental policies has consistently remained above 50% since 2010 (OECD, 2023). However, the key question remains: Are citizens actively involved in these decisions, and do they feel their voices are truly heard by their government?

4.3 Governance and Government

The concept of environmental governance, applied to the management of natural resources, can manifest in various ways (see Fig. 4.1). Environmental governance involves two key components: firstly, it refers to the process through which stakeholders with diverse interests organize themselves to establish a shared vision and make decisions related to environmental management. Secondly, it encompasses the set of processes and standards that regulate interactions between these groups of actors (Barriga et al., 2007). The term "stakeholder" has been used in organizational management since the 1960s and encompasses various definitions. In the context of environmental management, we adopt the definition of a stakeholder as a "group or individual who can impact or be impacted by the achievement of institutional objectives" (translated to English by the author, Mullenbach, 2007: 3).

Governance that arises from a negotiation process involving stakeholders is considered participatory. However, since citizens and nongovernmental groups have

[2] The shoreline protection law restricts the construction of buildings exceeding three storeys in height along the coast and delineates restricted areas beginning from the point where the sand ends (Cabrera, 2008; Montero, 2022).

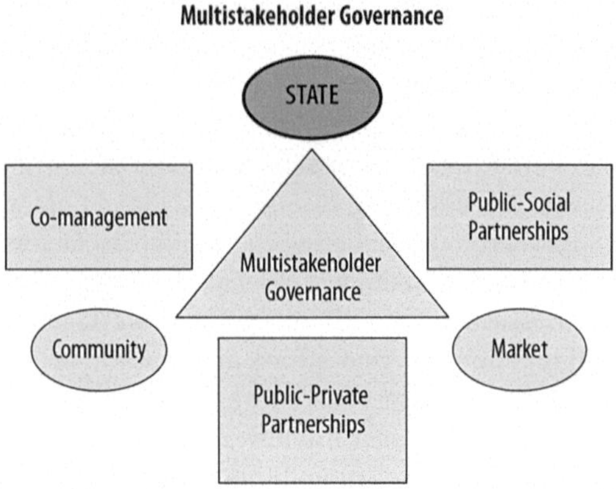

Fig. 4.1 Multistakeholder governance. (According to Agrawal and Lemos (2007) in Sasseville, 2010)

Table 4.1 Typology of comanagement

Typology of co-management		
7	Community control, partnership	Institutionalized joint decision-making; partnership of equals, where resources are managed locally, power delegated to the community
6	Management boards	Acting on common objectives; community is involved in policy-making as well as decision-making
5	Advisory committees	Partnership in decision-making starts; search for common objectives
4	Communication	Two-way communication starts; local concerns begin to enter research agendas
3	Cooperation	Use of local knowledge and native research assistants
2	Consultation	Face-to-face contact starts, community consulted about projects, feedback of research results
1	Information	Users are informed about rules and regulations.

Source: Adapted from Berkes (1994) and Arnstein's ladder of citizen participation (1969, cited in Berkes et al., 1991), Gravel, 2023: 477.

not always been included in decision-making processes, environmental governance scholars have emphasized the term "participatory" to denote processes that adhere to democratic and sustainable principles. Thus, participatory governance ensures that at least the first two levels of Arnstein's comanagement ladder are attained (see Table 4.1). Participatory governance, as outlined by Theys (2003), encompasses four key dimensions that serve as preconditions for its development:

> The modernization of public action, the establishment of horizontal coordination mechanisms, the broadening of traditional representations of rationality, and the redistribution of powers and roles among the state, other local and international institutions, the market, and civil society (in Gravel & Lavoie, 2009: 10).

Therefore, the processes that shape various forms of governance provide an opportunity to reassess the relationships between different groups of actors engaged in environmental management, highlighting the desirability of fostering partnerships among them.

However, there is a gap between the discourse promoted by advocates of sustainable forest management and the reality experienced when implementing such a model, as decentralization is not always a guarantee of equitable power redistribution among actors. In participatory governance, the involvement of local populations in the management process does not necessarily equate to gaining control over resource management (Sayer et al., 2005). Moreover, even when decentralization enables genuine participation of the local population in decision-making processes, some groups may be excluded or may feel inadequately prepared to participate meaningfully in the decision-making process. The first victims of this exclusion are often women, youth, indigenous peoples, migrants, and those from the most disadvantaged social classes (Gravel, 2006; Larson et al., 2006). It is crucial to consider the upstream stage of participation, which aims at capacity-building for historically marginalized groups within power circles to instill confidence and enhance the effectiveness of their intervention.

The limited emphasis on this preliminary stage is explained by divergent perceptions regarding the expected role of participants: as spectators or collaborators (Pretty, 1995). For some, participation is conceived in terms of utility and efficiency; it is desirable to ensure the success of the exchange between the professional team and local actors, assuming that the latter will be more inclined to accept the proposed changes if they are involved (spectator role). For others, participation is considered a fundamental right, allowing the initiation of capacity-building, collective action, and institutionalization among the targeted populations (collaborator role). Depending on the adopted vision, popular participation can be perceived as a tool facilitating the work of professionals or as an essential element in building human and social "capital" among participants (Sasseville, 2010).

These differing perceptions of participation leave room for interpretation, especially since participation can manifest itself in a multitude of forms. Berkes (1994) presented the typology of comanagement, which consists of seven levels ranging from simply informing the population about decisions already made to community control and the institutionalization of joint decision-making, illustrating the inequalities in the distribution of power between the parties, often to the detriment of the local population (Table 4.1).

The use of the term "participation," or the formulations used to denote variations in its application (e.g., comanagement, partnership, etc.), is delicate. In a study on partnerships in the field of sustainable forest management in Latin America, Ros-Tonen et al. (2007) highlighted the trap surrounding the use of such a term, as a positive connotation is often hastily attributed to it. This implicit value given to the alliance may risk masking inequalities among actors and, to a certain extent, even burnish the reputation of the minority holding power, displaying their good intentions to include other parties in the process (Sasseville, 2010).

In the progression outlined above, genuine participation begins to take shape at stage 3, where cooperation is established between decision-makers and community members. However, it is at level 5 that participation becomes increasingly tangible, as common objectives are identified through negotiation, fostering a more equitable relationship. Moving forward, levels 6 and 7 represent the realization of consensual decision-making, as community members gain membership on advisory bodies. Level 7, in particular, is seldom attained and serves as evidence of successful decentralization, as power is ceded to communities or advisory bodies, a departure from the State's typical regulatory role.

4.4 Methodological Overview

In order to examine the practical actions, outcomes, and challenges encountered by forest governance partners in Costa Rica, the Reventazón Model Forest (RMF) was selected as the study site. This "intersectoral social platform" (Rojas et al., 2020: 4) facilitates forest landscape management in Cartago province. In 2010, Master's student Émilie Sasseville under my supervision at Laval University conducted a field survey in the Reventazón Model Forest in May and June 2010. Along with document review and field observation, she conducted 16 interviews using a structured questionnaire with both management team members and Model Forest associates. Sasseville's research sheds light on the dynamics among stakeholders during the period following the creation of this first Model Forest in Costa Rica. It aims to elucidate the extent of local participation and evaluate its perceived significance by key stakeholders. Interviews with associates included questions about their organization's involvement in the RMF and their personal views on local participation and its impact on social capital. Triangulation, or cross-validation of information, was used to ensure data quality. These results provide insights into the early years of the Model Forest, which continues to evolve as a dynamic platform bringing together various groups.

4.5 What Governance for Costa Rica's Forests?

Typically, nonprofit organizations or governments, often from creditor nations, work with debtor nations to restructure debt payments. In return, the debtor nation allocates funds or resources toward initiatives that enhance environmental conservation, such as protecting natural habitats, promoting biodiversity, or supporting sustainable resource management. This approach aims to address both financial and environmental challenges, fostering a win–win situation for the debtor nation and the global environment (Idem). A debt-for-nature swap is an approach where a portion of a country's debt is forgiven or restructured in exchange for commitments to invest in environmental conservation or sustainable development projects (Isla, 2015). This innovative approach, initiated in 1987 through public–private partnerships and bond sales on foreign markets, facilitated the establishment of the first generation of national parks in Costa Rica (Idem; Gravel, 2023).

A subsequent generation of national parks saw the day in 1995, resulting from agreements between debtor and creditor governments, led to the emergence of what

Ana Isla terms "green capitalism" (Idem). These initiatives, complemented by programs addressing poverty, ecotourism development, and biodiversity management, have actively engaged local populations, including indigenous communities, women, and youth, in the productive aspects of protected areas (Idem). Costa Rica's commitment extends to offering services to international visitors through immersive experiences in its national parks, allowing them to explore the diverse tropical landscapes the country has to offer. The selection of potential candidates in such swaps involves careful consideration of natural amenities and political stability as key criteria (Deacon & Murphy, 1997).

As explained in an earlier publication (Gravel, 2023), this process, unfolding in a highly centralized nation where natural resource management was primarily overseen by the Ministry of Environment, Energy and Technology (MINAET) (Ferroukhi & Aguilar Schramm, 2003), resulted in a significant shift of territorial management authority. Costa Rica had to share control over its territory with international and local NGOs, acting as intermediaries between foreign governments (e.g., the United States, Canada, the Netherlands) and the Costa Rican government, leading to a dilution of national power within its own borders. While these measures reduced Costa Rica's external debt by 50%, they came at the cost of relinquishing decisive authority over crucial areas of its territory (Isla, 2015; Deacon & Murphy, 1997).

In an effort to elucidate socio-environmental injustices stemming from multistate cooperative agreements, Isla's (2015) study on the Arenal-Tilarán National Conservation Park underscores a management approach grounded in anthropocentric principles. This approach views the marginalization of indigenous communities as the adverse externalities resulting from the commodification of the natural world. According to Isla (2015), the pursuit of capitalist accumulation through "green capitalism" contributed to the erosion of indigenous communities' subsistence livelihoods, the erosion of traditional knowledge, and a disruption in the relationship between women and the environment. Nevertheless, acknowledging the intrinsic constraints of these swaps, Deacon and Murphy (1997) contended that the most likely alternative to such arrangements would involve engaging in agricultural activities, potentially focusing on monoculture plantations such as bananas or pineapples.

The Costa Rican government opted for environmental initiatives to enhance the sustainability of its territory and embraced the debt-swap proposition, making it one of the pioneers in achieving debt reduction through this innovative method (Isla, 2015). This diplomatic choice contributes to bolstering Costa Rica's global image as an environmentally conscious nation, positioning it as an international hub for ecotourism. Typically encompassing three or four entities, whether public or private, including foreign nations and both international and national NGOs, these multistate cooperation agreements involve the participation of federal power representatives. This strategic involvement ensures that these representatives are not directly linked to the presidency, a precautionary measure aimed at avoiding instability during political transitions (Deacon & Murphy, 1997).

However, within the country, there is a desire from various stakeholders affected by the management of natural resources to reconsider the distribution of power and responsibilities so that everyone can be heard in decision-making processes, which are aimed to be more horizontal. There is an increasing recognition of the advantage of opening debates to a variety of actors on issues related to sustainable development, acknowledging that no single entity possesses all the capabilities to manage environmental issues (Agrawal & Lemos, 2007).

4.6 The Model Forest, a Social Innovation in Costa Rica?

The International Model Forest Network[3] introduces an innovative forest landscape governance model driven by a voluntary association of people and institutions wishing to engage in the development of a common vision to manage their nature-based resources and forest landscape (Rojas et al., 2020: 4).

The methodology of the Model Forest (MF) emerged from the Canadian government during the 1992 Earth Summit in response to mounting conflicts over industrial forest management within the nation. Initially piloted in ten Canadian regions, model forests aim to foster collaboration among diverse stakeholders in the forestry sector, including governments, Indigenous communities, environmental organizations, forest workers, and local communities, to address sustainability challenges in forest landscapes (Ibid.). This innovative approach gained traction globally, leading to the establishment of model forests in approximately 35 countries worldwide. Despite their diversity, all model forests share common guiding principles, reflecting a cohesive network. Model forest management teams tailor their strategies to the unique characteristics of their territories, emphasizing long-term sustainability over short-term projects.

Established in 2002, the Latin American Model Forest Network (LAMFN) is a sub-continental alliance bolstered by an extensive array of private and public stakeholders, including governments, institutions, and international research organizations. Spanning across 13 countries[4], LAMFN comprises 33 model forests, making it the largest regional network of its kind globally (LAMFN website; Rojas et al., 2020). Its central hub is located at CATIE, the Tropical Agricultural Research and Higher Education Center, situated in Costa Rica's Cartago region.

Given the complexities of managing natural areas within a province, balancing the diverse interests of numerous stakeholders, the consultation platform of the Reventazón Model Forest Alliance (ABOMORE) emerged in 2003 as a valuable approach for consensual resource management in the Cartago region, east of San José. It encourages genuine public involvement from all levels of stakeholders—local, regional, and national—and decentralizes decision-making from centralized authorities. Representatives from the MINAET, which oversees protected areas in Costa Rica through the National Conservation Areas System (SINAC), participate

[3] The International Model Forests Network comprises 60 model forests across 35 countries (Red Internacional de Bosques Modelo website, 2023).

[4] For a complete list, see IMFN website at https://imfn.net/about/regional-networks/ibero-american-model-forest-network/

in these discussions. Decisions are made collectively by representatives from various protected areas entities according to the following principles (IMFN, 2019: 10):

- *Broad-based partnership*: Encouraging collaboration among various social sectors with vested interests.
- *Large landscape*: Focusing efforts on a diverse landscape representing various ecosystems and values.
- *Commitment to sustainability*: Promoting sustainable development, particularly in conserving and responsibly using forest ecosystems to maintain ecosystem services.
- *Good governance*: Establishing participatory governance structures with input from forest-engaged sectors, ensuring transparency.
- *Broad program of activities*: Designing programs that meet the needs and values of stakeholders involved in resource management.
- *Commitment to knowledge sharing, capacity building, and networking*: Prioritizing stakeholder capacity enhancement for sustainable resource management, fostering learning and networking for shared outcomes.

One question arises and will guide the upcoming discussion: In a nation like Costa Rica, known for its centralized governance, how is power distributed between the Reventazón Model Forest Alliance (ABOMORE) and the Costa Rican Government? The following case study will attempt to provide an answer and explain how Model Forests help reimagine forest governance.

4.7 The Territory of the Reventazón Model Forest

Encompassing an expanse of 312,000 hectares, the territory of the RMF stretches across the entirety of the Cartago province, renowned for its rich biodiversity and abundance of endemic species. Spanning elevations from 1300 to 11,000 feet (400 to 3500 meters) above sea level and experiencing annual precipitation ranging from 59 to 275 inches (1500 mm to 7000 mm), the region boasts a variety of ecosystems. The RMF's territory includes numerous public lands and private protected areas, which are integrated into seven national parks and the La Amistad and Cordillera Volcánica Central Biosphere Reserves. Additionally, two national wildlife corridors extend into the RMF's territory: the Central Talamanca Volcanic Biological Corridor (CBVCT) and the interurban Reventado Agua Caliente Biological Corridor (Cobri-SURAC) (see Fig. 4.2) (Jiménez, 2011).

The region where the Reventazón Model Forest is situated holds significant importance for Costa Rica's economy due to its proximity to densely populated areas, hydroelectric potential, abundant water sources, and the most fertile soil of the nation (IMFN website). It represents a diverse mix of interests and complex development dynamics. The population is varied, including individuals from the industrial zone of Cartago, laborers, farmers, landowners, entrepreneurs, and indigenous community members. The Cabécar ethnic group, known for its cultural

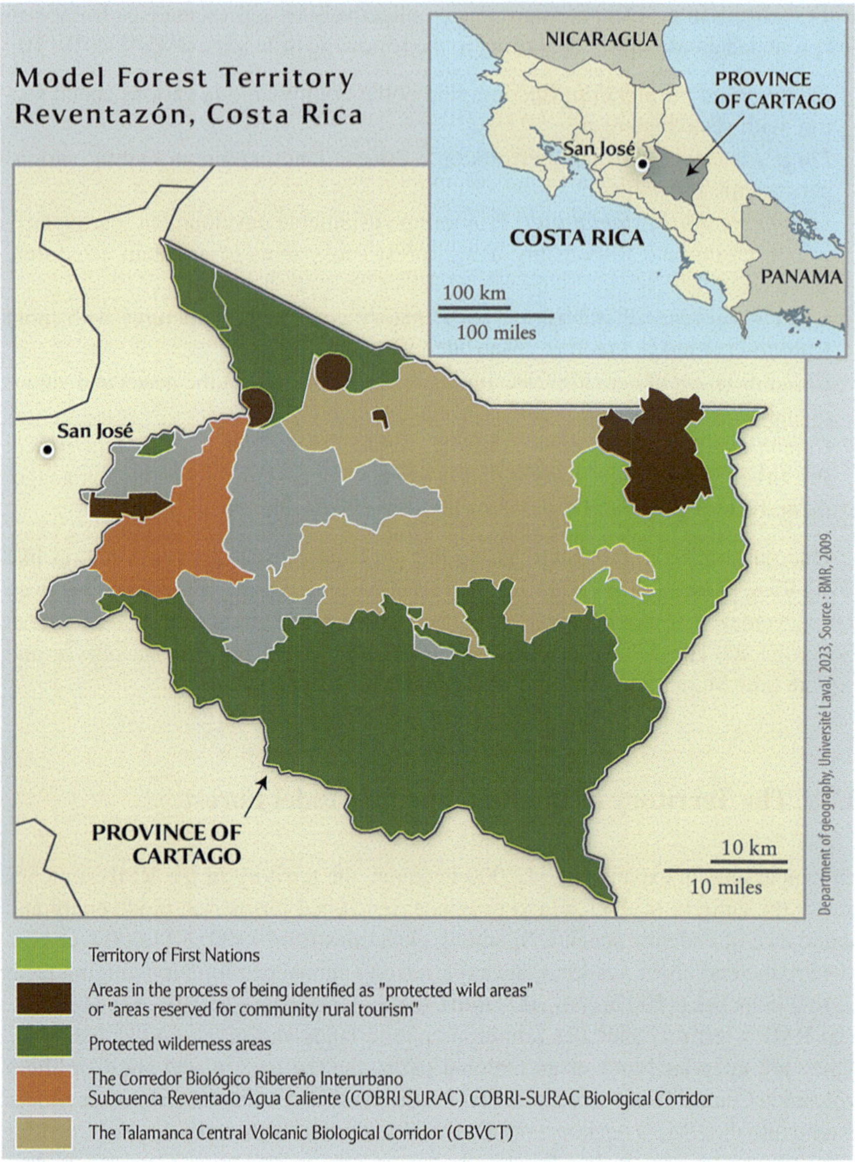

Fig. 4.2 The Reventazón Model Forest, Costa Rica. (Inspired by the IMFN website)

diversity, resides in three reserves partially located within the RMF territory, totaling 6000 members (Idem; Gravel, 2023).

This region encompasses not only areas experiencing environmental degradation and some of the best-preserved pristine primary forests but also thriving agricultural production—the region contributes to about 80% of the national vegetable production—vibrant communities, and marginalized indigenous reserves (Idem).

4.8 Building Networks: The Impact of Reventazón Model Forest

As per the members and associates interviewed, the Reventazón Model Forest primarily serves as a platform for communication and networking among stakeholders. It enables the sharing of experiences and learning in natural resource management in the region. The Model Forest brings together various stakeholders, including the private sector, public sector, NGOs, research and educational institutions, and citizen and indigenous groups, promoting horizontal relationships among them. The organization's strength largely lies in the alliances formed among these diverse actors, a benefit recognized by its members and associates.

Local partnerships are plentiful and varied. For example, the University of Costa Rica collaborates with local communities on reforestation projects using analogue forestry[5] techniques. Another instance is the meeting between the UNDP's SGP[6] team and leaders of local initiatives to fund sustainable development projects. The Model Forest management team's substantial coordination efforts to foster collaboration among its partners have been praised by numerous interviewees, including members and associates (Sasseville, 2010).

The RMF has successfully engaged stakeholders who previously weren't involved in sustainable resource management efforts in the region. For instance, Holcim S.A., a cement manufacturing plant in Agua Caliente de Cartago, became involved in RMF activities. During a visit to a Holcim plant in Argentina, the RMF team demonstrated the similarities between the company's sustainable development vision and that of a Model Forest, highlighting potential benefits of collaboration (RMF, 2010). This alignment suggests future collaborations, especially in environmental education. There's potential for the RMF to assist in rehabilitating abandoned extraction sites through environmental education and citizen involvement in regeneration tasks. In some cases, companies' cooperation with local communities is motivated by certification process requirements.

[5] Analogue forestry is an innovative approach to restoring damaged ecosystems. It aims to rebuild the structure and ecological balance of the original ecosystem while also introducing new species as needed (García Azuero et al., 2005; Sasseville, 2010).

[6] The Small Grants Program (Programa de pequeñas donaciones—PPD), funded by the Global Environment Facility (GEF) and managed by UNDP in Costa Rica, offers financial and technical support to community initiatives promoting environmental restoration and conservation (UNDP website). Since the Model Forest's establishment, over 15 initiatives in the region have received funding from the SGP, with support from the Model Forest organization (Durán Gárate, 2010). These projects mainly focus on community-based rural tourism and biological corridor management. To ensure measurable environmental outcomes, only projects within the boundaries of biological corridors are eligible for SGP funding.

4.9 Local Participation and Inclusion of Marginalized Groups

The RMF fosters collaboration across diverse sectors and aims to amplify voices often marginalized. Roger Villalobos, a member of the RMF's leadership, notes that without targeting specific groups, but focusing on citizen participation overall, the RMF has successfully engaged indigenous people, women, and youth in activities such as tourism, sports, and environmental initiatives (Sasseville, 2010).

Women and youth play active roles in managing biological corridors and sub-corridors, engaging in popular environmental education activities like bird watching, cycling competitions, and river cleaning campaigns in Turrialba (Jiménez, 2011). SGP-funded projects also attract their participation, such as ASOTUR in the Irazú Volcano region. ASOTUR, led by young people, mainly women, provides tourist services in the area. With support from the RMF management committee, ASOTUR received SGP funding for an environmental education and reforestation program in the region (Idem).

ASOTUR has become a key partner of the Forest Model, collaborating on tourism and environmental education projects with the SGP. This partnership has enabled ASOTUR, based in the Irazú region, to engage in various activities and training sessions in the province through RMF. It has also facilitated networking opportunities with other Forest Model partners, including the University of Costa Rica, which offers internships to ASOTUR students (Idem).

Furthermore, at the time of the study, it was noteworthy that 80–85% of the projects supported by the SGP in the Cartago area were submitted by organizations, half of whose members were women. The remaining projects, accounting for 15–20% of all projects funded by this program in the province, were administered by organizations whose members are exclusively women (Idem).

An official from the Sistema Nacional de Áreas de Conservación (National System of Conservation Areas) (SINAC) and a member of the management and direction committees of the Model Forest emphasizes the significant commitment of women to the RMF, with women strongly identifying with environmental themes. The participation of women from the region in governance processes can have a positive impact on the intergenerational transmission of values related to resource protection, thereby ensuring the continuation of long-term conservation goals (Idem).

As for the Jameikari indigenous community, they have been involved in establishing an ecotourism project in collaboration with the Model Forest in Manitoba, Canada. This alliance with the community is presented by the management team and the leadership of the RMF as one of their great achievements, as collaboration of this scale with indigenous communities is a first in the region (Idem).

In 2010, Sasseville's research shed light on an intriguing dynamic within the decision-making framework of the RMF. At that time, it was observed that the RMF featured an all-female management team, which stood in contrast to a predominantly male executive team. This gender composition of the leadership roles not only highlighted the diversity within the RMF but also brought attention to the role of women in environmental governance.

The public face of the RMF was further shaped by the prominent presence of women, notably exemplified by its manager at the time, Mrs. Mildred Jiménez. Despite the historical allocation of decision-making power primarily to the executive team, who were responsible for approving plans proposed by the management team, a nuanced balance was evident in practice.

This relationship between the two teams was characterized by a high degree of mutual trust and cooperation, establishing a horizontal dynamic within the RMF's governance structure. As a result, the management committee enjoyed considerable autonomy and freedom of action, effectively wielding decision-making authority on par with the executive team. Rather than imposing directives, the executive team provided guidance and support, fostering a collaborative approach to decision-making within the RMF (Jiménez, 2011).

Regarding the associates, numerous women actively engage with the RMF by overseeing the management of biological corridors[7], frequently comprising up to half of the members within the management committees. This trend was evident in Sasseville's study, where seven out of eight interviewed women held such positions. Additionally, several interviewees underscored the substantial involvement of young individuals in RMF-related activities, including their participation in various tourism projects.

Until 2010, indigenous communities were not represented in the RMF's executive or management bodies, despite management's encouragement. Logistical challenges, such as travel distance to meeting sites and limited availability, hindered their participation. These factors also affected the representation of local organizations, as quarterly meetings were held primarily in Cartago city[8], and participation was voluntary. This situation posed a limitation to both indigenous and remote residents' involvement in the RMF.

The management team aims to promote the involvement of marginalized groups in board activities and acknowledges this challenge. The decision to hold meetings in Cartago is based on the majority of board members residing in or near the city. RMF leadership is committed to enhancing diversity within its team and has begun efforts to include the Jameikari community in the steering committee, reducing their need for physical attendance at all meetings (Durán Gárate et al., 2021: 46).

"The board members recognize the great cultural value of this group and believe they can learn a lot from their knowledge of forest management, as indigenous communities in Costa Rica manage their natural resources on their territories," reported Roger Villalobos to Sasseville in 2010.

Since its establishment, the RMF has aimed to enhance cooperation among existing processes, reinforcing and replicating actions for participatory territorial

[7] Biological corridors, designated for biodiversity conservation, connect natural zones like Biosphere reserves, facilitating fauna migration and mitigating habitat fragmentation (Jiménez, 2011).

[8] Traveling to Cartago city can take several hours for residents of remote communities within the Reventazón Model Forest, requiring coordination with public transportation, especially in less well-served areas of the province.

management. Despite the increasing visibility of RMF's contributions, this platform still faced challenges and lacked recognition among the local population, as reported by several interviewed members and associates. Diffusion emerged as a weak aspect of the organization, indicating the need for further efforts to ensure its adoption by associates (Sasseville, 2010).

4.10 Fostering Shared Power Through Horizontal Organizational Structures

Can we say power is shifting among Costa Rican stakeholders in managing the forest landscape? When studying environmental governance, we analyze how power is shared among stakeholders, looking for signs of decentralization by the State or other influential actors. Talbot (2006) and Baron (2003) argued that power in this context is defined by interactions among stakeholders, not by certain groups or domination. While power dynamics can be uneven, with stakeholders often having different resources, it's important to recognize the complexity of these relationships.

In the case of the Reventazón Model Forest, where horizontal relationships are favored among partners, it is challenging to perceive power imbalances in these interactions. Instead, the strength of a common power is recognized, emerging from the union of voices. The coalition among different stakeholders allows them to gain greater visibility and wield more significant political influence.

In Costa Rica, land conservation policies encompass protected areas, indigenous territories, and lands owned by numerous private and corporate entities. Amid this complexity, the Model Forest Alliance has emerged as a valuable complementary strategy. By bringing together stakeholders involved in forest conservation, the Model Forest Alliance enhances overall conservation efforts in Costa Rica, fostering collaboration and synergy among different land management approaches. While Model Forests promote collaboration with different levels of government, we cannot directly say there is a shift in power. Rather, it is more about local involvement in development: the people of Cartago and regional stakeholders share responsibility for shaping their collective future alongside the state. Costa Rican authorities demonstrate signs of collaboration by appointing representatives from various ministries to serve on boards. This comanagement, reaching level 6 on Arnstein's ladder as presented by Berkes in 1994 (Level 6: "Management boards"; Acting on common objectives), signifies a notable advancement in granting autonomy to local population groups and establishing horizontal decision-making practices, as illustrated in Table 4.1.

4.11 Analyzing the Findings: A Comprehensive Discussion

To better grasp governance, let's rethink power. Some scholars, like Giddens (1990), suggest power isn't about centralized control but rather emerges from the actions of individuals or groups. While power is often seen as force or hierarchy, some authors like Allen (1999) view it as collaboration and collective action. In other words, power is not just about domination but also about working together (Talbot, 2006; Baron, 2003, and Sasseville, 2010).

In this view, social interactions can create power. According to Allen (1999), when communities combine resources like money, skills, or information, they generate power. Arendt (in Allen, 1999) stresses the importance of this collaborative effort, suggesting it's more significant than material factors in building power.

Through the RMF, collaborators in natural resource management in Cartago province share their knowledge, time, equipment, and ideas to achieve goals together, creating their own ability to act. This pooling of resources also builds human and social capital, as seen with a park ranger: "By participating in training and workshops, I gain, so the park gains, so the biological corridor gains. With these training sessions, my thinking shapes and evolves" (translated from Spanish by Sasseville, 2010).

While in many cases, environmental governance principles and collaboration efforts often only lead to the creation of discussion spaces (Lardon et al., 2008), the RMF appears to do more. Alliances between partners, actions taken, and involvement of marginalized groups show progress. Social and economic development advances are evident: rural community tourism thrives, local environmental initiatives grow, and citizens engage more in managing biological corridors. Among the RMF's contributions are funding waste management, conducting awareness workshops, educating youth on the environment, and carrying out reforestation activities (Sasseville, 2010).

The RMF greatly contributes to building social capital within communities by organizing and supporting activities, workshops, and training for citizens and organizations in the region. Having responsible, informed, and active participants is crucial for sustainable territorial development (Lardon et al., 2008, in Sasseville, 2010). Access to information and knowledge for all stakeholders is also vital for effective environmental governance processes (Beaurain, 2003). Furthermore, Pretty (1995) notes that limited efforts to involve and empower local populations in development projects often result in insufficient resources to maintain structures, activities, and practices after external support ends.

4.12 Conclusions and Prospects

The study of the emergence of the Reventazón Model Forest of Costa Rica has demonstrated that various stakeholders who depend on natural ecosystems to support their activities, and who are impacted by neighboring managerial actions, aspire to reevaluate the distribution of power and responsibilities through consensual, inclusive, and decentralized forms of decision-making. Furthermore, there is a growing recognition of the benefits of fostering dialogues on sustainable development issues, acknowledging that no single entity possesses all the capabilities required for effective environmental management (Agrawal & Lemos, 2007; Sasseville, 2010).

Despite the prevailing trend of centralized natural resource management in Costa Rica, the establishment of a model forest in Cartago Province has empowered local communities and stakeholders, enabling them to contribute to regional development. The establishment and methodology of the Reventazón Model Forest have revolutionized forest landscape management in the Cartago region and throughout Costa Rica. This initiative marked the first concerted effort to form an alliance involving all stakeholders engaged in forest protection. The ABOMORE Alliance, functioning as a platform, fosters collaboration among the government, private sector, and local community groups. This collaborative management approach, reaching level 7—Community control; partnerships: institutionalized joint decision-making, partnership of equals, where resources are managed locally, power is delegated to the community—represents the highest level of autonomy granted to local communities within a national territory on Arnstein's ladder of citizen participation (1969, cited in Berkes et al., 1991).

The Latin American Model Forest Network, hosted by the Tropical Agricultural Research and Higher Education Center (CATIE) in the town of Turrialba, Cartago (García Azuero et al., 2005), drives collaborative initiatives and provides vital technical support for sustainable development. The network's commendable efforts to amplify the voices of traditionally marginalized groups in Latin American resource management are noteworthy. For instance, the RMF has built strong relationships with Jameikari indigenous communities, actively involving them in various activities, including cultural tourism led by young indigenous guides. Recognizing the expertise of indigenous communities, who have long lived in and relied on forests, their integration into the RMF offers valuable insights and knowledge to participants (Durán Gárate et al., 2021).

In RMF governance, there's a focus on horizontal relationships among actors, fostering shared power from the collective voices of involved parties. Creating a coalition among forest-oriented interest groups boosts visibility and influence in decision-making at local, regional, and national levels. This collaborative approach aims to combine the strengths and resources of all stakeholders, turning the vision of sustainable development into real initiatives in the region (Idem).

Structures like the Reventazón Forest Model create good opportunities for participatory governance and sharing experiences. They could help fix issues pointed out by Isla (2015) and Deacon and Murphy (1997) in debt-for-nature agreements. Isla (2015) suggested these agreements often didn't consider the interests of local or indigenous groups, overlooking their cultures while following the agendas of Western-driven international organizations. It is important to note that despite their imperfections, these agreements spurred ecotourism in Costa Rica and contributed to the recovery of the country's biodiversity[9] (Robalino & Pfaff, 2013; Le Coq et al., 2012; UN Climate Change Website; Gravel, 2023).

[9]Alongside measures such as the 1995 Environmental Law, the 1996 Forestry Law introducing payments for ecosystem services (PES), and the 1998 Biodiversity Law.

References

Agrawal, A., & Lemos, M. C. (2007). A Greener Revolution in the making? Environmental governance in the 21st century. *Environment, 49*(5), 37–45.

Allen, J. (1999). Spatial Assemblages of Power: From Domination to Empowerment, dans D. Massey, J. Allen et P. Sarre (eds.) *Human Geography Today*, Polity Press, Cambridge, Royaume-Uni, 194–218.

Baron, C. (2003). La gouvernance : débats autour d'un concept polysémique. *Droit et société., 2*(54), 329–349. Retrieved at: https://doi.org/10.3917/drs.054.0329. Accessed 19 Jan 2024.

Barriga, M., Campos, J. J., Corrales, O. M., & Prins, C. (2007). *Gobernanza ambiental, adaptativa y colaborativa en bosques modelo, cuencas hidrográficas y corredores biológicos. Diez experiencias en cinco países latinoamericanos* (Technical document, no. 358). CATIE.

Beaurain, C. (2003). Gouvernance environnementale locale et comportements économiques, Développement durable et territoires [En ligne], Dossier 2. Accessed August 27, 2024. https://doi.org/10.4000/developpementdurable.1110

Berkes, F. (1994). Comanagement: Bridging the two solitudes. *Northern Perspectives, 22*(2–3), 18–20.

Berkes, F., George, P., & Preston, R. J. (1991). *Co-management: The evolution in theory and practice of the joint administration of living resources* (TASO Research Report, Second Series, no.1), 44 pages. Retrieved at: http://dlc.dlib.indiana.edu/dlc/handle/10535/1506. Accessed 19 Jan 2024.

Blaser, J., Küchli, C., Pierce Colfer, C. J., & Capistrano, D. (2005). Introduction. Background to a global exchange. In C. J. Pierce Colfer & D. Capistrano (Eds.), *The politics of decentralization: Forests, power and people* (pp. 1–9). Earthscan.

Cabrera, J. (2008). *Legal and institutional framework related to coastal tourism development in Costa Rica.* Center for Responsible Travel, Stanford University. https://icsid.worldbank.org/sites/default/files/parties_publications/C3164/Claimants'%20Expert%20Reports%20and%20Opinions/140425%20FTI%20ER1%20and%20Exhibits/FTI%2023%20__Legal_and_Institutional_Framework_Related_to_Coastal_Tourism_Development.pdf. Accessed 23 Apr 2024.

Capistrano, D., & Pierce Colfer, C. J. (2005). Decentralization: Issues, lessons and reflections. In C. J. Pierce Colfer & D. Capistrano (Eds.), *The politics of decentralization. Forests, people and power* (pp. 296–313). Earthscan.

Deacon, R. T., & Murphy, P. (1997). The structure of an environmental transaction: The debt-for-nature swap. *Land Economics, 73*(1), 1–24. Retrieved at: https://doi.org/10.2307/3147074. Accessed 19 Jan 2024.

Durán Gárate, L. (2010). Estrategia y mecanismos para la gobernanza de los recursos naturales en los Bosques Modelo Prince Albert (Canadá), Reventazón (Costa Rica) y Araucarias del Alto Malleco (Chile). Master's dissertation, unpublished, CATIE, Turrialba, Costa Rica.

Durán Gárate, L., Ruiz-Guevara, N., Villalobos Soto, R., & Gambetta, F. C. (2021). *La red latinoamericana de bosques modelo: pasado, presente y futuro* (Serie Tecnica, Technological Bulletin. n° 113, publication n° 25). CATIE. Retrieved at: https://repositorio.catie.ac.cr/bitstream/handle/11554/11486/La_red_latinoamericana_de_bosques_modelo.pdf?sequence=1&isAllowed=y. Accessed 19 Jan 2024.

Ferroukhi, L., & Aguilar Schramm, A. (2003). Avances y desafíos de la gestión forestal municipal en Costa Rica. In L. Ferroukhi (Ed.), *La Gestión forestal municipal en América Latina* (pp. 179–212). CIFOR, CRDI.

García Azuero, A. F., Campos, J. J., & Villalobos, R. (2005). Lineamientos para la planificación del Bosque Modelo Reventazón, Costa Rica. *Recursos Naturales y Ambiente, 45*, 74–80.

Giddens, A. (1990). The Consequences of Modernity, Polity Press and Basil Blackwell, Oxford, United Kingdom.

Gravel, N. (2006) (published in 2008). Les exclus du boom migratoire mexicain. *Cahiers des Amériques latines, 53*(3), 173–190. Retrieved at: https://journals.openedition.org/cal/1651. Accessed 19 Jan 2024.

Gravel, N. (2023). Antecedentes y trayectorias de la gobernanza participativa de los recursos naturales: lecciones de Canadá, Costa Rica y Brasil. In G. Massardier & V. Ugalde (Eds.), *Analizar la nueva fábrica de las políticas públicas: pluralización, territorialización, participación* (pp. 471–512). El Colegio de México.

Gravel, N., & Lavoie, A. (2009). Introduction. Environmental governance in perspective. *Canadian Journal of Latin American and Caribbean Studies. Special issue on environmental governance in Latin America, 34*(68), 5–18. Retrieved at: https://www.jstor.org/stable/i40084749. Accessed 19 Jan 2024.

International Forest Model Network (IFMN). (2019). *The international forest model network. A global approach to ecosystem sustainability.* Retrieved at: https://imfn.net/wp-content/uploads/2019/01/IMFN_Book_Eng_web.pdf Accessed 19 Jan 2024.

Isla, A. (2015). *The « Greening » of Costa Rica. Women, peasants, Indigenous Peoples, and the Remaking of Nature.* Toronto University Press.

Jiménez, M. (2011). *Concertación social para la gestión forestal: experiencia del Bosque Modelo Reventazón.* Retrieved at: http://www.cnf.org.pe/secretaria_conflat/memorias/DOCUMENTO%20MESAS/MESA%205/Mildred%20Jim%E9nez%20M%E9ndez.pdf. Accessed 19 Jan 2024.

Lardon, S., Tonneau, J.-S., Raymond, R., Chia, E., & Caron, P. (2008). Dispositifs de gouvernance territoriale durable en agriculture. Analyse de trois situations en France et au Brésil. *Norois, 209*(4), 17–36.

Larson, A., Pacheco, P., Toni, F., & Vallejo, M. (2006). *Exclusion e inclusion en la foresteria latinoamericana: hacia donde va la descentralizacion?* The Center for International Forestry Research.

Le Coq, J.-F., Pesche, D., Legrand, T., Froger, G., & Saenz Segura, F. (2012). La mise en politique des services environnementaux : la genèse du Programme de paiements pour services environnementaux au Costa Rica. *VertigO – la revue électronique en sciences de l'environnement, 12*(3), 2012. Retrieved at: https://journals.openedition.org/vertigo/12920, Accessed January 19, 2024.

Montero, J. (2022, October 14). The Costa Rica Maritime Zone and Concession Property. *The Tico Times.* Retrieved at: https://ticotimes.net/2022/10/14/the-costa-rica-maritime-zone-rica-and-concession-property. Accessed 23 Apr 2024.

Mullenbach, A. (2007). L'apport de la théorie des parties prenantes à la modélisation de la responsabilité sociétale des entreprises. *La Revue des Sciences de Gestion, 223*(1), 109–120. https://doi.org/10.3917/rsg.223.0109. Accessed 23 Apr 2024.

OECD. (2023). OECD *Environmental Performance Reviews: Costa Rica.* OECD iLibrary. Accessed August 27, 2024. https://www.oecd-ilibrary.org/sites/ec94fd4e-en/1/3/2/index.html?itemId=/content/publication/ec94fd4e-en&_csp_=6c66715l4db424ac0919ce9c979c1eec&itemIGO=oecd&itemContentType=book

Pretty, J. N. (1995). Participatory learning for sustainable agriculture. *World Development, 23*(8), 1247–1263.

Robalino, J., & Pfaff, A. (2013). Ecopayments and deforestation in Costa Rica: A nationwide analysis of PSA's initial years. *Land Economics, 89*(3), 432–448. Retrieved at: https://le.uwpress.org/content/89/3/432. Accessed 19 Jan 2024.

Rojas, C., Carrera, F., Imbach, A., Villalobos, R., & Durán, L. (2020). *Impacts and lessons learned from participatory landscape management platforms. The case of latin American model forests.* Latin American Model Forests Network. Retrieved at: https://imfn.net/wp-content/uploads/2021/08/RLABM-Impact-report-2020-FINAL-ENG.pdf. Accessed 21 Apr 2024.

Ros-Tonen, M. A. F., Van den Hombergh, H., & Zoomers, A. (2007). Partnerships for sustainable forest and tree resource management in Latin America: The new road towards successful forest governance? In M. A. F. Ros-Tonen (Ed.), *Partnerships in sustainable forest resource management: Learning from Latin America* (CEDLA, Latin America Studies, no 94) (pp. 3–35). Brill.

Sasseville, É. (2010). L'expérience de cogestion de la Forêt Modèle Reventazón, Costa Rica, vue par ses acteurs. Master's. dissertation, unpublished. Department of Geography. Université Laval.

Sayer, J., Elliott, C., Barrow, E., Gretzinger, S., Maginnis, S., Mcshane, T., & Shepherd, G. (2005). Chapter 6: Implications for biodiversity conservation of decentralized forest resources management. In P. Colfer & D. Capistrano (Eds.), *The politics of decentralization. Forests, people and power* (pp. 121–137). Earthscan Publications.

Talbot, D. (2006). La gouvernance locale, une forme de développement local et durable? Une illustration par les pays. Développement durable et territoires. Issue 7: *Proximité et environnement*. Retrieved at: http://developpementdurable.revues.org/index2666.html. Accessed 19 Jan 2024.

Theys, J. (2003). La gouvernance, entre innovation et impuissance. Le cas de l'environnement. Développement durable et territoires. Issue 2: *Gouvernance locale et Développement Durable*. Retrieved at: http://developpementdurable.revues.org/index1523.html. Accessed 19 Jan 2024.

Websites

AIDA Americas.org. (2019). *In historic decree, Costa Rica legally protects its corals*. https://aida-americas.org/en/press/in-historic-decree-costa-rica-legally-protects-its-corals. Accessed 21 Apr 2024.

International Forest Model Network (IFMN). https://imfn.net/. Accessed 21 Apr 2024.

Latin American Model Forests Network (LAMFN). https://imfn.net/about/regional-networks/ibero-american-model-forest-network/. Accessed 21 Apr 2024.

Reventazon Model Forest (RMF). (2010). https://imfn.net/regional-networks/reventazon-model-forest/. Accessed 21 Apr 2024.

United Nations Climate Change. Payments for Environmental Services Program/Costa Rica. https://unfccc.int/climate-action/momentum-for-change/financing-for-climate-friendly-investment/payments-for-environmental-services-program. Accessed 20 Jan 2024.

United Nations Development Program (UNDP) Small Grants Programme. https://www.undp.org/china/projects/small-grants-programme. Accessed 20 Jan 2024.

Chapter 5
Architects of Water-Smart Cities: An Actor-Mapping Approach

Abstract This chapter examines the Canadian water network's multi-scalar dimensions and its role in shaping water sustainability knowledge to support decision-making within Canadian communities. Originating from collaborative efforts in the POLIS Project at the University of Victoria, the water soft path advocates for a paradigm shift in water management, transitioning from a supply-centric to a demand-focused approach. The water resources co-management journey in Guelph, Ontario, is thoroughly examined, encompassing local citizen engagement, NGO participation, municipal water scientists, and interactions with watershed and provincial authorities. The narrative highlights the interconnected flow of information and proposals that have shaped Guelph's decision-making waterscape, fostering a commitment to water-smart landscaping and rewilding initiatives and positioning it as one of the pioneering water-smart cities in Canada.

Keywords Canadian water network · Co-management of water resources · Golden Horseshoe · Guelph, Ontario · POLIS Project · Rewilding initiatives · Water management · Water soft path · Water sustainability knowledge · Water-smart cities · Water-wise landscaping · Xeriscaping

5.1 Introduction[1]

This is the narrative of how my understanding of the Canadian water network began to take shape, resembling a patchwork of knowledge that mirrors its composition, the actors involved, their communication channels, and the various tools at their disposal, both tangible and intangible. It is a tale of knowledge in

[1]An earlier version of this chapter has been published as GRAVEL, N. and A. KONÉ (2017) The Guelph Water Connection: The Contribution of Actor-Network Theory (ANT) to the Study of Water Management in Guelph, Ontario. *Les Cahiers de géographie du Québec*. Special issue on water in Canada and Brazil. Ed. Louis Guay and Ana Britto. Vol. 61 (174): 489–512.

N. Gravel, *Tending Nature*, https://doi.org/10.1007/978-3-030-47062-3_5

motion—traveling, polymorphic, and multidimensional in nature[2]. While the commendable endeavors of some municipalities to foster water-smart cities[3] in Canada are noteworthy, they are part of a vast, interconnected network of entities and differentiated origins. This network, extending beyond provincial borders, thrives through its connections to research centers affiliated with universities and engages municipal government departments and nonprofit organizations across the country. The following provides a concise overview of the role of the water soft path approach (as introduced in this book) and its importance as it became the foundation for designing water-smart communities in Canada, particularly in Southern Ontario.

5.2 Context

Upon my arrival at the University of Guelph as a visiting professor during my sabbatical leave in 2010–2011, a serendipitous encounter in the bookstore led me to a profound discovery—perhaps the most significant of the year, if not the decade, for a researcher like me. While exploring the natural resources book section, I stumbled upon a book titled "Making the Most of the Water We Have: The Soft Path Approach to Water Management," published in 2009 by authors David B. Brooks, Oliver M. Brandes, and Stephen Gurman from the University of Victoria, British Columbia. Intrigued by the "soft path approach," I eagerly delved into the contents, hopeful that it might unveil a theoretical framework I had long sought. Indeed, it did. By focusing on reducing water demand rather than increasing supply, the approach highlights and values the strategies employed by inhabitants of drought-ridden regions.

Many parched societies, such as those in Islamic or Mediterranean regions or any human group facing semiarid conditions, have mastered the art of water management by: (1) reducing water uptake; (2) maximizing water use by reusing graywater in the kitchen or for carwashes, choosing water fit for specific uses; and/or (3) desalinating or treating water for further use (Damman et al., 2023; Malloy, 2021). While freshwaters were long perceived as infinite in countries with abundant volumes like Canada, the United States, and Brazil, the industrial age has revealed the scarcity of unpolluted waters. Acidic rains in the 1980s and algal blooms in remote areas shocked many, emphasizing that large quantities of freshwater don't necessarily mean an abundance of quality freshwater.

[2] For an example of the application of ANT see Gravel and Koné, 2017.

[3] "Water-smart cities" refer to urban areas that employ advanced technologies, data-driven strategies, and sustainable practices to effectively manage, conserve, and optimize water resources. These cities focus on intelligent water management systems, incorporating technologies like sensors, data analytics, and real-time monitoring to enhance water efficiency, reduce wastage, and ensure the sustainable use of water. The aim is to create urban environments where water resources are managed intelligently, promoting conservation and resilience in the face of challenges such as population growth, climate change, and water scarcity (Damman et al., 2023).

While companies are often blamed as major users of water resources, the impacts of municipalities and residences drawing, using, wasting, and releasing water into nearby water bodies must be considered. Superficial runoff waters from the streets carry deposits from vehicle emissions, chemicals, de-icing salts, lawn pesticides, and street litter. The destination of this water, rarely treated before its release, is crucial. Untreated polluted water depletes the total drinking water reserve available for human and nonhuman beings and burdens aquatic ecosystems, often out of sight, such as invisible oceanic life.

My deep interest in water demand reduction, explored through the prism of the soft water path, prompted me to forge connections with themes of environmental citizenship and public participation that I had been delving into in previous years. When invited to address a sociology seminar at the University of Guelph in 2010, I introduced the soft water path as a seemingly novel methodology. My colleague Tony Winson posed a crucial question: "What about the hard path?" I elucidated that the "hard path approach" managed the water offer, rather than the demand, focusing on water mains, pipes, and water-retaining infrastructures, emphasizing the delivery of water where needed without questioning the necessity of constructing new infrastructure. The consequence was an excessive investment in built infrastructures, carrying economic, environmental, and societal externalities.

5.3 Charting New Waters: Pioneering the Water Soft Path Approach

In response to the limits shown by the prevailing water management approach and the luring climate change crisis, researchers in environmental sciences, activists, and a growing number of concerned municipal staff in Southern Canada championed the water soft path approach. This shift was catalyzed by experiences across the country where municipalities, faced with the impending need to reassess their water infrastructures and the costs of replacement, questioned the status quo: "Do we really need this much water?" This pivotal moment coincided with the formation of the POLIS research group on water management at the University of Victoria. Researchers such as David B. Brooks, Oliver M. Brandes, and Stephen Gurman (2009) embarked on studying water demands under three scenarios: no change, a 15% reduction in demand, and a 30% reduction in demand. The outcomes of this initiative, coupled with infrastructure upgrades, the promotion of eco-friendly appliances in homes and workplaces, and thorough checks on water losses from transportation pipes, have allowed some municipalities to contemplate the future. Despite growing population trends, they no longer need to plan for new water surpluses.

Upon reaching out to the POLIS team and presenting my work on water governance in Latin America, an unexpected turn of events occurred: I received an invitation to share my findings during one of their weekly seminars in 2012. Combining this opportunity with plans to attend a conference in Kelowna, I made my way to

Vancouver Island. The well-attended seminar resonated with the audience, as we shared a common interest in water governance, albeit with slightly different perspectives.

Given my background in critical social theory, my focus lies in understanding the actors, their dynamics, and distribution of power in water governance. Specifically, I explored how citizens and nonprofit groups could influence and participate in decision-making processes related to water allocation, as well as engage in policing, surveying, and auditing practices. In contrast, the POLIS staff brought an engineering twist to my perspective. They had expertise in green technologies, enabling water audits, and the examination of water supplies and losses in various buildings and institutions.

From their viewpoint, citizen participation in water management primarily involved end uses, such as adopting water-saving devices and practices by replacing faucets and toilets and minimizing daily indoor and outdoor water usage. The approach also included offering incentives, such as rebates and subsidies, to help citizens make their residential spaces more environmentally friendly. For instance, in the city of Victoria, citizens were encouraged to exchange old shower heads for new eco-friendly ones at no cost.

However, this shift in practices seemed to be driven primarily by municipal governments informed by scientific research, rather than grassroots citizens' groups. The challenge lies in mobilizing voluntary participation in efficient water consumption as a catalyst for change. While some water-saving initiatives contribute to the resource protection, these programs should be accompanied by public awareness and education campaigns to enhance coherence among actions and deepen understanding of the ecological processes and reasons behind such efforts. In this way, citizens' ecological awareness could give rise to grassroots social innovations.

In preparation for my presentation to the POLIS team, I delved into the work of the POLIS network on Salt Spring Island, one of the Gulf Islands off the east coast of Vancouver Island in British Columbia. It is the largest and most densely populated of the Gulf Islands. A study led by Maas and Porter-Bopp (2010) outlined water management scenarios for the island, which had a population of 11,635 in 2022, reflecting a 10.2% increase since 2016 (Salt Spring Exchange, 2022). This population growth was attributed to the island's appeal to retired Canadians seeking the country's mildest climate, thanks to its position on the leeward side of Vancouver Island.

Given my expertise in semiarid environments, I recognized Salt Spring Island as an ideal case study in water governance, and I decided to take the ferry boat from Vancouver Island to reach it. During my stay on SSI, I gathered information on local water-related challenges. On the final day, I discovered that the owner of the B&B where I stayed was the islands' director, who provided me with valuable insights during a conversation over locally produced rosé!

The SSI director highlighted the challenges of water allocation, stemming from aging infrastructures and insufficient funds for replacement. The small size of the municipalities on the island exacerbated the situation. In some cases, 500 residents had to share the substantial cost of renovating wells and pipe infrastructures, a consequence of past administrations overlooking the need for financial planning. The absence of municipal incorporation further complicated matters, as the island received directives from the Capital Regional District in Victoria, the Capital City of British Columbia.

My host, the director, faced the daunting task of addressing these challenges. The lack of a local governance structure on Salt Spring Island meant that residents were distant from decision-making centers and politically underrepresented. The geographical remoteness of places like Salt Spring Island from centers of power makes it difficult for residents to influence decisions that significantly impact them. Conversely, distant decision-makers do not always grasp the territorial challenges, and many lack insights regarding fund allocation.

After my talk in Victoria, Oliver Brooks, the POLIS Project manager, mentioned his recent collaborations in Ontario and provided me with a list of contacts, including Mr. Wayne Galliher, the then director of water installations in the city of Guelph. I had a sense of coming full circle. The list also led me to a collaborator in the Kitchener-Waterloo region, whom I contacted later. This person recommended reaching out to two others: a water planner from the city of Guelph and another from the Peel region, known for its fusion landscaping initiatives. Intrigued by the term "fusion landscaping," I continued my exploration journey in the Toronto suburb of Mississauga, which eventually led me to develop an interest in the Credit River Watershed and the Toronto Greenbelt.

5.4 Weaving Water Knowledge: Connecting the Dots

The meshing of the different components of the Canadian water network, encompassing know-what (information, facts), know-why (science, laws, principles), know-how (capabilities), and know-who (social skills), as outlined by Morgan and Murdoch (2000: 84), will be explored. Regardless of the type of knowledge, whether derived from a farmer's observations or a water technician's expertise, all facets need consideration in unraveling the network (Fabricius et al., 2006). Emphasis will be placed on the role of institutions, particularly municipal governing bodies, as gatherers and custodians of know-how and their relationship with the citizenry.

Crucial questions will be raised: Are municipal environmental programs alone sufficient to create a green city, or is public participation required, and to what extent? Is the involvement of citizens at every level of decision-making necessary, or is there another "invisible" ingredient that fosters a population's awareness of ecological limits and readiness to act within those boundaries?

The city of Guelph has emerged as a pioneer in green living and water conservation within Ontario, standing out as one of the cities with the highest quality of life in Canada. Its journey to this level of innovation, participation, and program implementation involved a multifaceted network of actors both within and beyond the city's borders.

In the findings of my research published in 2017, co-authored with graduate student Adama Koné, it became evident that Guelph's initiatives were backed by a diverse network of stakeholders across Canada. This network, operational since at least 2005, has played a vital role in equipping Guelph's staff with a steady stream of information, data, comparative studies, and state-of-the-art methods for implementing water-efficient measures and technologies.

The expansive nature of this network may not be readily apparent to the average city resident. However, certain salient points, such as local food fairs, university conferences, and workshops organized by collaborative initiatives like Transition Guelph, offer glimpses of the interconnectedness. These events convene city advisors with expertise in urban agriculture and outdoor water efficiency programs, NGO personnel, and specialists in green energy, water conservation techniques, and the preservation of wild pollinators.

The robustness and organizational prowess of the Guelph water network are derived from its diverse stakeholders. Supported by universities, research centers, environmental NGOs, and think tanks, Guelph's municipal council and service personnel are well-versed and current. Empowered by this knowledge, they advocate for water fee increases to a segment of the population resistant to change, promoting a proactive and informed stance toward sustainable practices.

5.5 Data Collection and Research Methodology

During my year-long leave as a professor at the University of Guelph in 2010–2011, I started this research employing a diverse data collection strategy. Methods included participatory observation, engagement in local environmental networks, conference attendance, visits to agricultural fairs, workshops, and gardens, daily observation of residents' gardening practices, and 15 interviews with city staff and leaders of environmental groups. Some of them agreed to give follow-up interviews as well in 2016. Additional interviews and on-site visits were conducted in the cities of Waterloo, Kitchener, and Toronto (Peel region), as well as on Salt Spring Island (BC). The central inquiry posed to interviewees was: "Where did the idea of water management in Guelph (or any city) come from?" Subsequent questions explored the influence of the water soft path approach on local decision-making, the primary partners in water management, and the contributions of other Canadian cities to raising awareness about water resource planning and management in Guelph.

This chapter employs actor mapping to unravel the intricate web of connections among different actors and understand the network's structure that underpins efficient knowledge production and communication. The Pan-Canadian water soft path connection elucidates the accumulation of expertise on water conservation and sustainability within various Guelph-based institutions, establishing the City of Guelph as a Canadian municipal leader in water conservation.

5.6 The Soft Path to Water Management

The soft path approach, conceived by Amory B. Lovins in the 1970s for addressing energy production and consumption challenges, found adaptation 30 years later in the realm of water through the work of Canadian water researchers Brandes and Brooks (Brooks et al., 2009) within the POLIS Project on water sustainability (see Sect. 2.9 of this book). This approach advocates treating water as a service, prioritizing ecological sustainability, delivering water of appropriate quality, and planning from the future back to the present (Idem). In contrast to conventional water demand management, the water soft path considers carrying capacity and long-term collective objectives. It encourages communities to envision scenarios over the next 15, 30, and 50 years, facilitating collective reflection on their desired future. These scenarios include a status quo, illustrating potential resource scarcity, and brighter alternatives achieved through demand reduction. Subsequently, the community collaboratively devises a water resource management plan aligned with its values, potentially embracing strategies like a "zero-growth policy" focused on efficient resource use and reduced consumption (Brandes & Brooks, 2006).

How can a city, facing population growth pressure, plan to utilize existing water sources without adding new infrastructure while meeting the future needs of its residents? The answer lies in the hands of governing bodies, particularly municipal governments, which can take on a leadership role in water education through campaigns, activities, events, and programs, implementing a water soft path approach. Engaging citizens, developers, and business owners through a blend of awareness-building initiatives, restrictive regulations, and fines, while also actively listening to the needs of the community members, emerges as indispensable. Achieving this delicate balance is challenging, but in Guelph, Ontario, a unique combination of thriving civic culture and environmental awareness has made it possible.

5.7 Water-Smart Cities

Water-smart cities represent an innovative and sustainable urban development paradigm that integrates advanced technologies, policies, and practices to efficiently manage and optimize water resources within an urban environment. This concept aligns with the principles behind the water soft path and the broader framework of smart cities, which leverage digital technologies and data-driven approaches to enhance the quality of urban life, address environmental challenges, and promote resource efficiency.

In the context of water-smart cities, the emphasis is particularly on the intelligent and adaptable management of water-related systems and services to achieve urban sustainability and resilience. This includes the integration of smart water meters, sensor networks, and real-time data analytics to monitor and optimize water distribution, consumption, and quality. Water-smart cities strive to bolster water resilience, minimize wastage, and promote sustainable water utilization within the urban landscape. This approach places a significant emphasis on supporting ecosystem functions, ensuring the provision of ecosystem services, and prioritizing the availability of an adequate water supply to sustain ecological diversity, encompassing both vegetation and animals, particularly during periods of drought (Damman et al., 2023).

Water-smart cities represent a forward-looking and adaptive approach to urban water management, aiming to ensure sustainable water resources for growing urban populations while addressing environmental concerns and enhancing overall urban resilience (Idem; Malloy, 2021).

5.8 Water Management and Population Growth in Southern Ontario: A Case Study

The City of Guelph, located on the outskirts of the Greater Golden Horseshoe (Fig. 5.1) to the west of the Toronto Greenbelt, confronts water supply challenges stemming from biophysical constraints. These include a restricted fresh water supply, strained aquifers, and the looming risk of drought years (O'Flanagan, 2015). Designated as a residential development possibility under the Growth Plan Act, Guelph faces significant water provisioning constraints. As of 2021, the city's population stands at 145,000, making it one of the fastest-growing cities in Canada with a growth rate of 9.1% between 2021 and 2016, surpassing both the national (5.2%) and provincial (5.8%) growth rates for the same period (Binstock, 2010: 7; Guelph Today, 2022). These conditions have heightened the collective urgency for sustainable resource management. The Ontario government's 2006 growth plan, setting growth limits for Toronto, spurred Guelph officials to prioritize sustainable water practices.

The Ontario government's inaugural Places to Grow Act (2005) outlined development and legislative planning for the subsequent 25 years in the Golden Horseshoe region (see Fig. 5.1), prioritizing the dispersal of housing away from Toronto, the province's largest city. By 2015, the Greater Toronto Area was home to 8 million people, projected to reach 14.8 million by 2051 (Government of Ontario, 2020, website). The 2005 initiative aimed to decentralize population growth and employment opportunities to surrounding areas, enhancing the quality of life in the metropolis. This approach also intended to address environmental conservation goals, including reducing car emissions, increasing green space within the city, minimizing commuting needs, and establishing vibrant, planned green neighborhoods. Protecting the Lake Ontario shoreline, one of the Great Lakes, was a significant concern, fostering collaborations between municipalities within its watershed. For instance, the Peel Region Climate Change Strategy (2011) resulted from collaboration among seven governing bodies and local ENGOs. As part of the collective effort to enhance Lake Ontario's resilience, one measure implemented was to

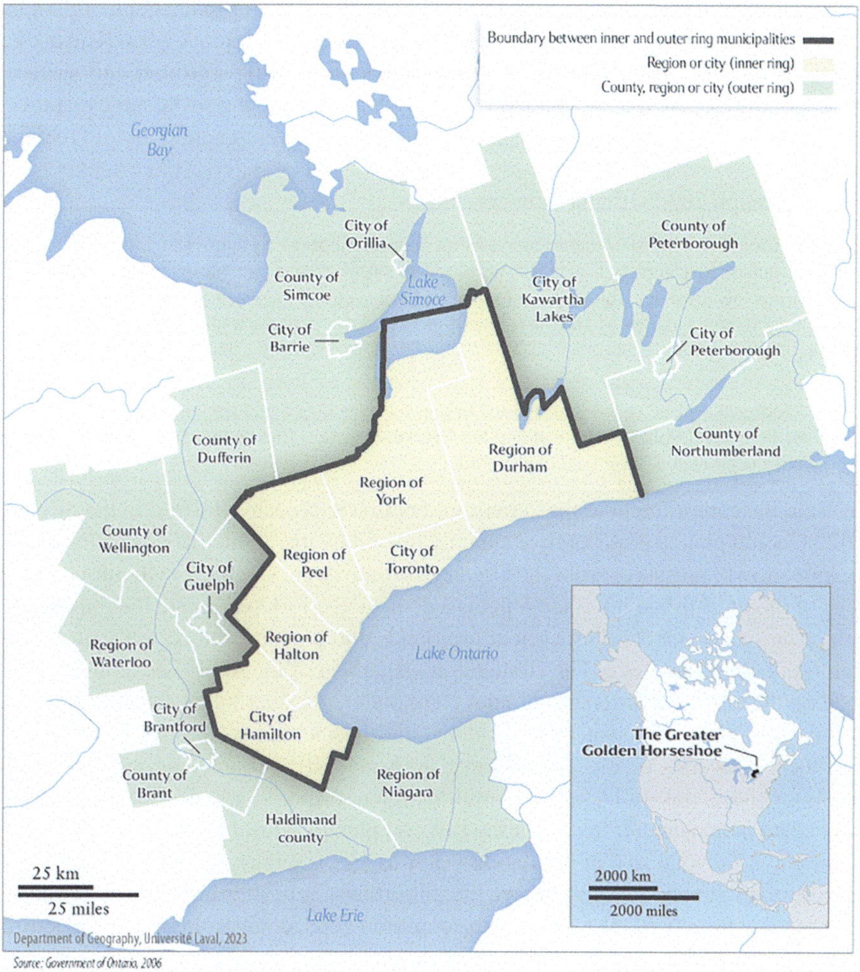

Fig. 5.1 The Greater Golden Horseshoe, Ontario, Canada

encourage municipalities that extract water from the lake to return treated water to it (Peel Region, 2011). Additional measures included replacing pavement with porous materials to mitigate heavy runoff, which contributes to shoreline erosion.

However, the Ontario Government's Places to Grow Act does not include any provisions for a drinking water supply policy. This is despite the reality of a burgeoning population with a high water-consumption rate—Canadians consume more than twice as much drinking water as Europeans! —which necessitates costly construction of new water facilities, including new sources, transportation, and treatment, along with associated environmental costs. Water provisioning is therefore handed over to the municipalities, while the plan only shows concern for protecting streams and head sources.

This narrative will explore how Guelph adopted a water co-management plan in 2005 and how citizen involvement, with a participation rate above 50% (Binstock, 2010), has been a driving force. The finite character of underground and surface water continues to feature prominently in city planning, guiding sustainable resource management and influencing decisions on residential development (City of Guelph, 2022: 8).

The Guelph 2022 Official Plan articulates:

> One of the City's most valuable assets is its Natural Heritage System. The City takes an environment first approach and is committed to protecting, maintaining, enhancing and restoring the diversity, function, linkages, and connectivity between and among natural heritage features and areas and surface and ground water features within the city over the long term.
>
> Development will be prohibited within defined features in accordance with the provisions of the Provincial Policy Statement and the Growth Plan [...] Ensure that water quality and quantity is protected, improved or restored.

Note the commitment to the "environment first approach" outlined in the official 2022 Guelph plan, a philosophy staunchly upheld by the municipal governing body despite growth pressures from the Greater Toronto Area. Guelph's meticulous planning at the rural-urban interface reflects a dedicated effort to mitigate the impacts of urbanization on surrounding agricultural lands and preserve the region's rural and agrarian cultural heritage. The strategic management of the Aberfoyle aquifer, situated on the city's outskirts, is a crucial component, considering its significance as a source of high-quality water for residents and businesses, including Nestlé Water and local breweries (Case, 2016; Gravel & Koné, 2017). Balancing the extractive needs of these stakeholders, even multinational corporations, necessitates robust watershed management. In navigating this complexity, a resolute municipal government plays a pivotal role in regulating water usage. The intricate dynamics of natural resources governance underscore the importance of reasserting the state's role to safeguard natural heritage assets for the common good, countering occasional shifts toward more liberal and neoliberal political ideologies at the provincial level. The Guelph water network, as presented here, serves as the linchpin for the city's overarching "environment first approach" as articulated in the 2022 official plan (City of Guelph, 2022).

5.9 Water Resources in Guelph: Challenges and Exploitation

Guelph faces unique challenges in its water resources, primarily relying on groundwater due to the scarcity of streams and significant rivers on its territory. The city is characterized by two rivers with consistently low water levels and discharge capacity issues. Within the broader Grand River watershed, which supplies 70% of drinking water from underground sources and 30% from surface waters, Guelph's dependence on groundwater is pronounced (Veale & Cooke, 2017: 380; Galliher, 2010).

Groundwater, however, is a finite and slow-renewing resource, compounded by the exploitation activities of beverage companies. Nestlé's Aberfoyle Springs plant, situated at the city's southeastern end, extracts water from the Gasport aquifer at an annual rate of 762 million liters as of 2015 (The Wellington Advertiser, 2016). Nestlé Waters Canada benefits from favorable regulations, allowing wholesale extraction of water, leading to controversies and opposition from local water activists for over two decades. Concerns revolve around potential environmental impacts on Aberfoyle Creek and local underground water, along with fears of losing sovereignty over the aquifer (The Wellington Advertiser, 2016; The Globe and Mail, 2016).

The Guelph area, with its limited water sources and variable annual precipitation, experiences semiarid conditions in certain years, particularly during hot summer seasons. Over the past 15 years, the region has faced challenges such as increased temperatures and notable dry periods. Drought years were recorded from 1997 to 2003, with additional dry spells in 2007 and 2011 (Clean Air Partnership, 2012: 9).

Despite its designation as having a humid continental climate, characterized by an average annual precipitation of 905 mm, the distribution of rainfall is concentrated between April and November. This distribution pattern contributes to significant variations in river flow within the watershed (The Grand River Watershed website). The combination of limited water availability and climate variability poses additional complexities for water resource management in the Guelph region.

5.10 A Comprehensive Actor Mapping Analysis

Guelph has garnered recognition within the province for its proactive stance on environmental issues. As identified by De Loe et al. (2002 in Binstock, 2010: 9) and Shrubsole et al. (2017), factors contributing to this environmental awareness include political commitment, citizen involvement, linkages with external agencies, and institutional arrangements. The authors note that Guelph's distinctive feature lies in its "continual and multidimensional" community involvement, emphasizing sustained engagement over time (Ibid.) The following section depicts the diverse network of actors engaged in water management in Guelph. Actor mapping represents the relationships and interactions among individuals or groups involved in this context, providing insights into their roles and connections within the broader framework.

Grassroots initiatives, originating from *water activist groups*, have played a pivotal role in shaping public opinion and promoting the conservation of natural and heritage assets in the Guelph region. These initiatives leverage effective communication through local media, family events such as the 2Rivers Festival, and conferences with local partners. Events like the 2Rivers Festival provide platforms for cross-pollination of ideas and networking, fostering exchanges among people with diverse allegiances.

One notable movement is the *Transition Guelph initiative*, which advocates for alternative energy sources and efficient strategies in areas like transportation and housing fixtures, contributing to Guelph's trajectory toward becoming a green city. Supporting these groups and initiatives is the *Guelph Civic League*, which not only fosters civic values and community engagement but also runs an in-school water education program with provincial funding.

Civil society and church groups, representing the "know-how" and "know-who" elements, contribute significantly to the community's strengths. A comprehensive study on water activism in Guelph identified a network of over 158 water activists, including members of the ENGO Wellington Water Watchers (WWW) (Case, 2016). WWW, along with the Canadian Council, has actively advocated for recognizing water as a common good. They have taken legal action against Nestlé Waters Canada to prevent water extraction from the local aquifer, particularly during drought periods, challenging the terms of Nestlé's permit (The Globe and Mail, 2016; Gravel & Koné, 2017). The collective efforts of these water activists underscore the importance of community-driven initiatives in safeguarding Guelph's water resources.

The City of Guelph Staff and Councilors. Guelph's journey toward environmental sustainability has seen significant shifts, particularly in the roles played by the City Council and its elected officials. In 1999, the City of Guelph initiated a water conservation and efficiency strategy, later reinforced by a Water Supply Master Plan in 2006 (subsequently updated in 2014 and 2019). However, during this period, the City Council faced challenges in implementing these measures, encountering resistance to even minor changes (Binstock, 2010).

A pivotal transformation occurred with the election of the City Council in 2009, which demonstrated a more favorable stance toward environmental innovation. The new council implemented robust communication strategies, leveraging a dynamic website with daily updates, e-newsletters, rebate programs for water-saving household fixtures, and recognition programs for outstanding environmental stewardship by citizens and business owners. This marked a significant departure from the previous approach, showcasing the importance of political dynamics in environmental governance.

Since 2006, *citizen engagement* has played a crucial role in reshaping the focus from traditional infrastructure-centric approaches (hard path) to a more nuanced emphasis on demand management (soft path). Initiatives like the Landscape Visit Program exemplify this shift, offering citizens the opportunity to consult with landscape assessment advisors. These advisors provide guidance on creating environmentally friendly front and backyards by adopting water-wise gardening practices, embracing organic methods, and installing rainwater harvesting systems. This collaborative approach underscores the importance of community involvement in steering Guelph toward a water resilient future (Fig. 5.2).

The Guelph Outside Water Use Program, established in 2002 in response to the Ontario Low Water Response Plan, is a notable success in conserving Guelph's

Fig. 5.2 Dry-climate landscaping in a Guelph neighborhood
Description: Dry-climate landscaping, or xeriscaping, in the front yard of a Guelph residence showing complete lawn removal and the utilization of porous material such as mulch and stones. (Photo: Gravel, 2012) Xeriscaping is a term coined in the 1970s in Denver, Colorado, combining the prefix "xeros" meaning dry, and the suffix "scape," meaning view. It refers to dry landscaping techniques that do not need extra irrigation water. This principle has become recognized in gardening practices and may be applicable beyond arid zones as well. Gardeners should opt for plants that thrive in accordance with the natural precipitation levels of their local area (Reed & Stibolt, 2018: 78)

groundwater supply and communicating information to citizens to mitigate drought impacts during hot, dry summer months (City of Guelph website, 2012, 2016; Government of Ontario, 2024 website). The program classifies water use into three levels based on weather conditions: Level 0 (Blue) for careful use, Level 1 (Yellow) for reduced outside use, and Level 2 (Red) for reducing and stopping nonessential use. Administered by the City of Guelph Water Services division and enforced by bylaw officers, it is explicitly designed as a "mitigation strategy, not a disaster relief or emergency response plan" (Roth & De Loë, 2015: 21). With over 95% compliance from Guelph citizens, the program yields impressive summer water savings of 8.6 million liters per day (Idem; Interview with Guelph Water Services staff, May 2012). The positive outcomes reflected the efforts of the Guelph water conservation network and increased the City Council's ability to influence the provincial government, leading to a revision of the population growth threshold within the Ontario Growth Management Plan (Clean Air Partnership, 2012). The successful water resource management in Guelph is underpinned by principles of negotiation and compromise, demonstrating flexibility to integrate new actors and initiatives, thereby ensuring the resilience of the network over time and adapting to changes in elected officials.

According to Veale (2015: 353), coordinator of Policy Planning and Partnerships for the Grand River Conservation Authority (GRCA), the governance model adopted for water management in Ontario empowers conservation authorities to play a crucial role in facilitating participatory approaches to enhance watershed health and resiliency. An illustrative instance is the collaboration with the Ontario government and other partners, securing funds over a 2-year period to develop the Grand River Water Management Plan, launched in March 2013. This plan aligns with the Ontario Water Opportunities Act passed in 2010, aimed at conserving water supplies, sustaining water infrastructure, and positioning Ontario as a North American leader in new technologies and services for water conservation and treatment.

Commencing in 2006, proactive measures were initiated to enhance the Grand River Watershed's resilience to climate change, recognizing the potential impacts of abrupt changes in rainfall patterns, termed extreme events. These events lead to unforeseen consequences such as increased flooding, surface erosion, and diminished fisheries, wetland, and marsh habitat in the basin and along the Lake Erie shores (Farwell et al., 2008). Climate change has become a unifying cause, bringing together environmentalists and the general population to collaborate. It has proven to be an effective catalyst for making cities "more blue" and reconsidering water resource management provisions in Canada (Guelph Water Services staff, interview, June 2012). In fulfilling these functions, the GRCA and other collaborative watershed stakeholders embody the region's strengths in terms of water conservation and transformative actions toward the establishment of "blue cities," respectful of their river ecosystems and water resources. Roth and De Loë (2015: 21) emphasize the strength of the Ontario approach, highlighting the "inclusion of provincial government and conservation authority representations on Water Response Teams."

5.11 Discussion: Water-Smart Planning and Regulation

Developing environmental commitment at a city scale necessitates municipal public policies that cultivate heightened environmental consciousness throughout society, fostering collective engagement that eventually encompasses all stakeholders. This is why the architects of water-smart cities should not only be urban planners, engineers, and hard science-trained staff but also educators, communicators, social workers, and policy analysts. Enhancing a community's resilience through the reduction of its water footprint, incorporating flexibility in water governance to address unforeseen challenges, and tailoring programs to meet the needs of individuals from diverse backgrounds are all imperative steps.

This is particularly true for water-saving programs based on a user-pay approach aimed at curbing water demand within watersheds and preserving scarce resources. In times of economic turmoil, when thousands of industrial jobs are lost in Canada, how are low-income households and charity organizations coping with high utility bills? Have city planners considered adopting a tiered system to set water delivery and treatment costs? Few Canadian cities have delved into the sociology of their users, as interviews with city water managers revealed in southern Ontario. In contrast, in Seattle, Washington, measures have been taken to adapt water tariffs to households' paying capacity. Some Canadian cities, like Cochrane, Alberta, have replaced their unique average cost pricing structure with a tiered system to curb excessive water consumption, with the lowest block rate having a fixed price immune to increases. This allows household members to manage their water usage and avoid being charged in the same proportions as heavier water consumers.

Most water efficiency policies were designed by municipal agencies when water stocks were perceived as declining, whether due to underground water depletion or surface water contamination. Some municipalities, faced with the necessary replacement of water infrastructure, are planning to charge for water, especially if residents have not contributed to special funds in previous years (as seen in the case of Salt Spring Island).

Attention should be directed toward the requirements of the most vulnerable population to actively involve them as willing contributors in their city's initiatives for resource conservation. If they are left with the bitter taste of rising utility bills (as seen sometimes in Guelph where citizens pay for drinking water and wastewater treatment), a segment of the population will not naturally be inclined to make more efforts to conserve other resources, thinking that they already do enough for the environment—a catchphrase often heard in Canadian cities.

It has been demonstrated that the most effective strategy to reduce water demand is to combine the user-pay approach with environmental education and outreach programs. The objective is to unite residents toward a common cause: the construction of sustainable cities:

> A Green City is a healthy, sustainable, livable city with a healthy natural environment, healthy people, and a thriving economy. A Green City has clean air, water, and soil. In a Green City, homes and businesses are nestled among abundant trees, natural areas, and greenspaces. Urban wildlife such as birds and frogs can also thrive in a Green City. (Credit Valley Conservation, 2010 brochure).

The climate change looming threat has been real enough to get different stakeholders to sit around the table and start planning for the future. The example of the Peel region, situated west of Toronto, is illustrative: Six governmental entities, including the region of Peel and three municipalities, collaborated with two conservation authorities to develop climate change strategies spanning the period from 2011 to 2050. Their comprehensive report assessed various aspects, including water conservation, water runoff management, urban forests, alternative energy technologies, construction norms for new buildings, and the urban transportation system. This thorough evaluation led to a reexamination and consolidation of existing green techniques and strategies, all framed within the overarching strategy of climate change mitigation.

In this context, the catalyzing force of impending climate change served to organize environmental conservation efforts and mobilize citizens in a manner that may not have occurred without the looming threats of warmer summers, water scarcity, and devastating flash floods. Coupled with the shared water source with the city of

Toronto, Lake Ontario, and the risks posed by its receding coastline and diminishing biodiversity, there are ample and compelling reasons to motivate a significant number of citizens to begin caring for their local environment. Whether it's the unnoticed water creek, overlooked marshes, or the looming expansion of industrial activities along streams, these factors prompt residents of the Peel region to take notice and take action. As residents became more engaged, they began supporting conservation authorities financially, actively participating in meetings and activities, and spreading awareness within their communities. The outcome is dynamic, well-structured organizations deeply rooted in their communities, drawing interest from individuals nationwide. Another advantage of initiating action is an increase in influence at the political level and the attraction of experts to serve on organization boards. This success is initially local but is beginning to extend its influence across a broader geographical scale.

5.12 Conclusions: Building a Multiscalar Network for Water Management

The research delves into the influence of the pan-Canadian network of water stakeholders on shaping sustainable water policies in Guelph, fostering a healthier and more resilient watershed basin in response to climate variability. Drawing insights from the POLIS Project contributors at the University of Victoria, who presented Water Soft Path applications in various Canadian cities, including Salt Spring Island, Kitchener, Elora (Maas and Porter-Bopp, 2011), and Guelph, the study unveils interconnected communication channels uniting academics and interest groups. It underscored the robust alliances between elected officials and citizens when shared visions align. The positive response to municipal water-saving initiatives demonstrates that the majority of Guelph residents agree with and are willing to contribute to sustainable water management.

As the investigation uncovers links between the university-led network and municipal spheres, the focus shifts to understanding how this Canadian water connection influenced decision-makers in Guelph. This exploration reveals a shared goal among involved parties and the existence of a human and physical network facilitating multi-scalar knowledge construction among water stakeholders from provincial governments, research centers, conservation authorities, and municipalities. The physical components of the system include modern information and communication technologies, such as web platforms, enabling swift flows and exchanges of information.

In the collaborative aftermath, the municipal Water Service staff in Guelph actively utilizes the shared water conservation knowledge from Canadian thinkers to formulate the Guelph Water Conservation and Efficiency Strategy Update (2009) (Gravel and Koné, 2017). Although the Water Soft Path approach was not explicitly mentioned in interviews with Guelph water managers and planners in 2012, its influence is clearly evident in the design of the Guelph Water Strategy. There is no doubt that they were aware of its existence, as stated in interviews and documented by Binstock in 2010. In his report's preamble, Binstock explicitly designates Guelph as a "pilot project" to identify enabling conditions that allowed the city to pursue measures aligned with the soft path's core principles.

Given that the network existed prior to the research, the study involved retracing the "spiderweb" thread in a retrospective process to understand the network's structure. This approach offers the advantage of observing how citizens have appropriated knowledge and their inclination to embrace water-wise practices.

Several indicators, including citizen participation in various city-organized activities, the voluntary conversion of lawns, and the presence of rain barrels on private lots, serve as valuable metrics for assessing the extent of citizen involvement. Over time, it becomes clear that collective knowledge has been shaped and shared, establishing a solid citizen base for the network. This foundation ensures the network's continuity beyond the tenure of elected officials and City staff, reflecting its robustness and sustainability.

The community's commitment to managing its shared water resources is succinctly captured by the statement, "Despite the addition of about 20,000 households, Guelph's water use has decreased twenty percent over the past ten years" (Idem). Follow-up interviews with City Water Services staff in August 2016 revealed that "Water conservation is also saving customers money due to the ability to delay major water infrastructure projects that would be needed to supply the City with more water." This approach keeps water rates low and affordable, as the City avoids the need for additional supply infrastructure (Idem).

In the context of water management in Guelph, actor mapping has proven invaluable in comprehending the interconnections within a system that evolves through diverse stakeholder contributions over time (Gravel and Koné, 2017). This approach provides an external and comprehensive perspective on the process of establishing a multi-scalar network of knowledge holders in water management. By examining the network's developmental context, it becomes clear how ideas disseminate through various channels (websites, e-newsletters, conferences, workshops, newspapers, studies, etc.), gain support from influential figures, and are adopted by interest groups accross various administrative bodies and at various stages.

References

Binstock, M. (2010). Moving toward a soft path approach? A case study of water management in Guelph, Ontario. *Canadian Institute for Environmental Law and Policy.* Retrieved at: http://cielap.org/pdf/towardsoftpathapproach.pdf. Accessed 13 Jan 2024.

Brandes, O., & Brooks, D. B. (2006, revised in 2007). *The soft path for water in a nutshell.* Victoria, B. C., POLIS Project, University of Victoria and Friends of the Earth Canada. Retrieved at: https://dspace.library.uvic.ca/bitstream/handle/1828/7963/The%20Soft%20Path%20for%20Water%20in%20a%20Nutshell.pdf?sequence=1&isAllowed=y. Accessed 13 Jan 2024.

Brooks, D. B., Brandes, O. M., & Gurman, S. (2009). *Making the most of the water we have. The soft path approach to water management*. Earthscan.

Case, R. A. (2016). Social work and the moral economy of water: Community-based activism and its implications for eco-social work. *Critical Social Work, 17*(2). Retrieved at: https://doi.org/10.22329/csw.v17i2.5903. Accessed 13 Jan 2024.

City of Guelph. (2012). *Outside water use program. Accessed June 26 2013.* http://guelph.ca/living/environment/water/waterconservation/outside-water-use/

Credit Valley Conservation. (2010). Brochure.

Damman, S., Schmuck, A., Oliveira, R., Koop, S. H. A., Almeida, M. d. C., Alegre, H., & Ugarelli, R. M. (2023). Towards a water-smart society: Progress in linking theory and practice. *Utilities Policy, 85*: Retrieved at: https://www.sciencedirect.com/science/article/pii/S0957178723001868?via%3Dihub. Accessed 13 Jan 2024.

Fabricius, C., Scholes, R., & Cundill, G. (2006). Mobilizing knowledge for integrated ecosystem assessments. In W. V. Reid, F. Berkes, T. J. Wilbanks, & D. Capistrano (Eds.), *Bridging scales and knowledge systems. Concepts and applications in ecosystem management* (pp. 165–182). Island Press.

Farwell, J., Boyd, D., & Ryan, T. (2008). *"Making watersheds more resilient to climate change". A response in the Grand River Watershed, Ontario Canada.* Grand River Conservation Authority. Retrieved at: http://www.grandriver.ca/AboutGrand/ClimateChange08.pdf, Inaccessible online.

Galliher, W. (2010). *City of Guelph. Water conservation and efficiency program.* Presented at the Canadian Water Summit. Retrieved at: http://fr.slideshare.net/CWS_2010/wayne-galliher-city-of-guelph-city-of-guelph-water-conservation-and-efficiency-program. Accessed 13 Jan 2024.

Government of Ontario. (2012b and 2024). Ministry of Natural Resources and Forestry. Surface Water Monitoring System. Accessed August 27, 2024. https://www.ontario.ca/page/low-water-response-program

Gravel, N. (2012). The Politics of Drought in the Brazilian Northeast. Paper presented at the Annual Conference of the Canadian Association of Geographers (CAG), Waterloo, ON.

Gravel, N. & Koné, A. (2017). The Guelph Water Connection: The Contribution of Actor-Network Theory (ANT) to the Study of Water Management in Guelph, Ontario. *Les Cahiers de géographie du Québec. Special issue on water in Canada and Brazil. Ed. Louis Guay and Ana Britto. 61(174): 489–512.* https://www.erudit.org/fr/revues/cgq/2017-v61-n174-cgq04102/1053663ar/

Maas, C., & Porter-Bopp, S. (2010). *A Soft Path Strategy for Salt Spring Island, BC* (POLIS Discussion series paper. POLIS project on Ecological Governance). University of Victoria, B.C., Environmental Law Center. 60 p. Retrieved at: https://poliswaterproject.org/polis-research-publication/soft-path-strategy-salt-spring-island-b-c/. Accessed 13 Jan 2024.

Maas, C., & Porter-Bopp, S. (2011). *A soft path strategy for Fergus-Elora, Ontario. A soft path for water case study* (POLIS Discussion Series Paper 11-01. Executive Summary), 10 p. Retrieved at: https://poliswaterproject.org/polis-research-publication/soft-path-strategy-fergus-elora-ontario/. Accessed 13 Jan 2024.

Malloy, C. (2021). How to build a water-smart city. *Bloomberg*. Retrieved at: https://www.bloomberg.com/news/features/2021-08-02/how-to-build-a-water-smart-city. Accessed 13 Jan 2024.

Ministry of Natural Resources, Canadian Government. Clean Air Partnership. (2012). Guelph water conservation programming. In *Accelerating adaptation in Canadian communities. Case studies.* Toronto. Retrieved at: http://www.cleanairpartnership.org/wp-content/uploads/2016/06/Accelerating-Adaptation-in-Canadian-Communities-2.pdf. Accessed 13 Jan 2024.

Morgan, K., & Murdoch, J. (2000). Organic vs Conventional Agriculture: knowledge, power and innovation in the food chain. *Geoforum, 31, 159–173.* https://doi.org/10.1016/S0016-7185(99)00029-9

Reed, S., & Stibolt, G. (2018). *Climate-wise landscaping. Practical actions for a sustainable future.* New Society Publishers.

Roth, A. P., & De Loë, R. C. (2015). *Addressing drought: A survey of Canadian and international experiences.* Water Policy and Governance Group. Retrieved at: http://www.wpgg.ca/sites/default/files/Roth%20and%20de%20Loe%202015.pdf, Inaccessible online.

Shrubsole, D., Waters, D., Veale, B., & Mitchell, B. (2017). Integrated water resources management in Canada: The experience of watershed agencies. *International Journal of Water Resources Development, 33*(3), 349–359. Retrieved at: https://ideas.repec.org/a/taf/cijwxx/v33y2017i3p349-359.html. Accessed 13 Jan 2024.

The Globe and Mail. (2016, August 21). *Nestlé continues to extract water from Ontario town despite severe drought: Activists.* http://www.theglobeandmail.com/news/national/nestle-continues-to-extract-water-from-ontario-town-despite-severe-drought-activists/article31480345/. Accessed 13 Jan 2024.

The Wellington Advertiser (2016). *Puslinch report suggests Nestlé operations impact local groundwater.* https://www.puslinchtoday.ca/2016/08/puslinch-report-suggests-nestle-operations-impact-local-groundwater/. Accessed 13 Jan 2024.

Veale, B. J. (2015). The conservation authority governance model: A unique case in Ontario, Canada. In L. Ruelas Monjardín, A. Travieso Bello, & O. Sánchez Sánchez (Eds.), *Gobernanza ambiental: teoría y práctica para la conservación y uso sustentable de los recursos* (pp. 329–359). Plaza y Valdés Editores. Retrieved at: https://pdfcoffee.com/2017-ruelas-monjardin-l-r-gonernanza-para-el-manejo-integral-pdf-pdf-free.html. Accessed 13 Jan 2024.

Veale, B., & Cooke, S. (2017). Implementing integrated water management: Illustrations from the Grand River watershed. *International Journal of Water Resources Development, 33*(3), 375–392. Retrieved at: https://www.tandfonline.com/doi/abs/10.1080/07900627.2016.1217503?journalCode=cijw20. Accessed 13 Jan 2024.

Websites

Canadian Heritage Rivers System, The Grand River winner of the Thiess International River Prize. https://chrs.ca/en/rivers/grand-river. Accessed 13 Jan 2024.

City of Guelph. Be Water Wise. http://guelph.ca/living/environment/water/water-conservation/be-water-wise/. Accessed 12 July 2013. Inaccessible online.

City of Guelph. *Water/wastewater rates.* http://guelph.ca/living/environment/water/water-rates/. Accessed 13 Jan 2024.

City of Guelph. *Outside water use program.* http://guelph.ca/living/environment/water/water-conservation/outside-water-use/. Accessed 13 Jan 2024.

City of Guelph. (August 11, 2016). *Guelph still at Level 2 Red for outside water use. City hallnewsroom. Accessed August 25, 2016.* http://guelph.ca/2016/08/guelph-level-2-red-outside-water-use/

City of Guelph. (2006, 2014, 2019). *Water supply master plan.* https://guelph.ca/plans-and-strategies/water-supply-master-plan/. Accessed 30 Oct 2023.

Government of Ontario. Ministry of Natural Resources and Forestry. *Surface Water Monitoring System.* https://www.ontario.ca/page/low-water-response-program. Accessed 13 Jan 2024.

Government of Ontario. *Conservation Authorities.* https://www.ontario.ca/page/conservation-authorities. Accessed 13 Jan 2024.

Government of Ontario. (2005). *Growth plan for the greater golden horseshoe region.* https://www.placestogrow.ca/index.php?option=com_content&task=view&id=9&Itemid=14. Accessed 13 Jan 2024.

Government of Ontario. (2020). *Ministry of municipal affairs. A place to grow: Growth plan for the greater golden horseshoe.* https://www.ontario.ca/document/growth-plan-greater-golden-horseshoe/where-and-how-grow. Accessed 30 Oct 2023.

Grand River Conservation Authority. (2013). *GRCA newsroom, June 24, 2013. Haldimand Landowners can get tips on improving natural areas.* http://www.grandriver.ca/Newsroom/News.cfm?id=721. Accessed 12 July 2013. Inaccessible online.

Grand River Conservation Authority. (2013a). *River data. GRCA Monitoring network- River conditions summary.* http://www.grandriver.ca/index/document.cfm?Sec=2&Sub1=10&Sub2=0. Accessed 12 July 2013. Inaccessible online.

Grand River Conservation Authority. (2013b). *River Data. River Flow, Water Quality and Weather Data.* http://www.grandriver.ca/index/document.cfm?Sec=2&Sub1=0&sub2=0. Accessed 12 July 2013. Inaccessible online.

Grand River Conservation Authority. (2012). *GRCA newsroom, January 16, 2012. Province supports water management plan with grant of 903, 000$.* http://www.grandriver.ca/Newsroom/News.cfm?id=632. Accessed 12 July 2013. Inaccessible online.

Grand River Conservation Authority. *Heritage River Designation.* https://www.grandriver.ca/en/our-watershed/Heritage-River-Designation.aspx. Accessed 13 Jan 2024.

Guelph Today. (2022). *Guelph's population grew by nearly 10 per cent from 2016 to 2021, census shows.* https://www.guelphtoday.com/local-news/guelphs-population-grew-by-nearly-10-per-cent-from-2016-to-2021-census-shows-5045325. Accessed 13 Jan 2024.

O'Flanagan, R. (2015). Water a Worry for Future Guelph. *The Guelph Mercury Tribune.* http://www.guelphmercury.com/news-story/5601383-water-a-worry-for-future-guelph/. Accessed 13 Jan 2024.

POLIS Project on Ecological Governance, University of Victoria, B. C. http://www.poliswaterproject.org/. Accessed 13 Jan 2024.

Region of Peel. (2012, 2020). *The Peel Region 2012 (2020) Water and Wastewater Master Plan.* https://www.peelregion.ca/public-works/environmental-assessments/2020-masterplan.asp. Accessed 13 Jan 2024.

Region of Peel. (2011). *Peel Region Climate Change Strategy.* https://www.peelregion.ca/planning/climatechange/reports/pdf/climate-chan-strat-bgr.pdf. Accessed 5 Sept 2016. Replaced by Peel2041. Regional Official Plan Review. Climate Change. Pelle 2042 Discussion Paper (2018) https://www.peelregion.ca/officialplan/review/pdf/climate-change-discussion-paper.pdf. Accessed 13 Jan 2024.

Rural Water Quality Program, Grand River Conservation Authority. https://www.grandriver.ca/en/our-watershed/Rural-Water-Quality-Program.aspx. Accessed 13 Jan 2024.

Salt Spring Exchange. (2022). https://saltspringexchange.com/2022/02/09/salt-spring-island-population-increases-10-2-between-2016-and-2021/. Accessed 13 Jan 2024.

Transition Guelph Movement. http://www.transitionguelph.org/. Accessed 12 Aug 2016. Inaccessible online. Replaced by https://www.facebook.com/TransitionGuelph/events/?ref=page_internal&locale=it_T&paipv=0&eav=Afata9tXfk7Ox8OEFO93SYfDyMO49vRCdwVoRGOBJQwYkg9bkam1vQbJ3z99gUnM1wM&_rdr

Waterbucket. The partnership for water sustainability in BC (2011) New water soft path strategy proposed for Ontario communities of Fergus and Elora. http://waterbucket.ca/wuc/2011/04/03/new-water-soft-path-strategy-proposed-for-the-ontario-communities-of-fergus-and-elora/. Accessed 13 Jan 2024.

Chapter 6
Unleashing the Flow: Dam Removal and Governance of Salmon Rivers in the State of Maine, USA

Perchance after a few thousand years, if the fishes will be patient, and pass their summer elsewhere, meanwhile, nature will have leveled the Billerica Dam and the Lowell factories, and the Grass-ground, the river runs clear again to be explored by new migratory shoals, even as far as the Hopkinton Pond and Westborough Swamp.

Henry David Thoreau, A week on the Concord and Merrimack Rivers.

Abstract This chapter examines the paradigm shift in values and priorities that paved the way for the gradual acceptance of dam removal on Maine's salmon rivers as of the 1990s. The changing perspective explored herein signifies a significant shift in the environmental consciousness of politicians, lawmakers, journalists, anglers, and the broader population of Maine. This evolution involves an increasing recognition of and preference for natural amenities over outdated historical hydroelectric infrastructures. Observing the decline of fish populations has profoundly impacted the public, fostering heightened awareness of the extensive degradation of river ecosystems beyond their sustainable capacity over the past century. This awareness has sparked a collective movement to restore fish habitats and cultivate a society and economy rooted in more environmentally friendly principles and values.

Keywords Actor mapping · Advocacy · Atlantic salmon · Anadromous fish · Dam removal · Environmental consciousness · Environmental alliances · Free-flowing rivers · Hydroelectricity · Kennebec river · Maine · Natural amenities · Penobscot river · Post-industrial paradigm shift · Restoration movement, River ecosystems · Salmon rivers · Values and priorities, Wabanaki tribes, Penobscot Nation

N. Gravel, *Tending Nature*, https://doi.org/10.1007/978-3-030-47062-3_6

6.1 Introduction[1]

The restoration of Atlantic salmon river habitat in Maine is inherently complex, involving a vast array of entities over the past 70 years. Alliances among river stakeholders—including influential Indigenous Wabanaki tribes, white water rafters, anglers, conservationists, private trusts, foundations, lawmakers, organizations, and governmental agencies at various geographical scales— were crucial in persuading the few remaining defenders of historical hydroelectric dams, which obstructed fish passage for minimal electricity production. Through collaborative efforts in research, advocacy, negotiation, education, and community outreach, a compelling river restoration narrative was crafted, significantly shaping the evolving interactions with hydroelectric power companies and local communities.

Engaging the Maine public and fostering an understanding of the potential role it could play in advocating for healthy rivers required both patience and time. Well-organized coalitions, in collaboration with the news media, effectively achieved their fundraising objectives. This collective effort facilitated the acquisition and removal of main stem dams on major salmon rivers, such as the Kennebec and Penobscot Rivers—marking a pioneering achievement in the United States.

I have chosen to explore the contrasting narratives surrounding fish river habitat restoration and dam removal versus dam maintenance and fishway construction in Maine during the latter half of the twentieth century. The laudable endeavors of civic coalitions along the Kennebec and Penobscot Rivers, marked by the successful demolition of main stem river dams commencing in 1999, subsequently emerged as a paradigmatic exemplar for the broader national free-flowing river movement. Despite an estimated 74,000 dams still existing in the United States, the country leads globally in dam removal, with hundreds of small- to medium-sized dams already dismantled. In 2017 alone, 86 dams were demolished, contributing to a total of approximately 1300 in recent years (International Rivers website).

Maine's influence has resonated across the nation, with numerous US states following its example. The pivotal moment was the denial of a license renewal for the Edwards Dam on the Kennebec River in Augusta by the Federal Energy Regulatory Commission (FERC), leading to its breach in 1999. The river flowed thereafter unobstructed for 57 miles, up to the Fort Halifax Dam, which was demolished in 2008 (Natural Resources Council of Maine website).

Other significant salmon rivers in the Northeast, formerly inhabited by substantial populations of wild Atlantic sea-run salmon, include the Connecticut, Merrimack, Saint Lawrence (Canada), Grand Cascapédia (Quebec, Canada), Restigouche (Quebec, Canada), and Saint John Rivers (binational US–Canada watershed). These river habitats, once abundant with Atlantic salmon (Salmo salar), have undergone environmental shifts, and commercial fishing is no longer permitted. Only sport fishing and Indigenous sustenance fishing are allowed (Kurlansky, 2020: 117; Lee, 2020).

[1]An earlier version of this chapter was presented the Annual Conference of the Canadian Association of Geographers (CAG), Winnipeg, May 27–31.

6.2 Methodological Considerations

Between 2018 and 2019, I held a visiting professorship in the Department of Environmental Studies at the University of New England, in Biddeford, Southern Maine. My research extended northward along the salmon rivers to the University of Maine in Orono and various regions across Maine, encompassing the South, the Lakes, MidCoast, North, and Downeast.

To explore the complex dynamics of the network supporting Atlantic salmon river habitat restoration in Maine, I established contact with associations and fish researchers at the Maine Sustainability and Water Conference in Augusta in March 2019. As a Canadian researcher, I sought a rapid immersion into the nuanced aspects of salmon river habitat restoration in Maine. The Water Conference proved instrumental, providing valuable insights into the efforts of diverse groups within this sphere and fostering connections with sea-run fish activists and organizations.

Throughout 12 field trips spanning from the Quebec province to the State of Maine, journeying along the Old Canada Road in the Kennebec River Valley, I gathered data via observations and conducted 15 semi-structured interviews with river and fish scientists and conservation experts. To maintain the precision and currency of information, follow-up interviews were conducted in 2022. These individuals graciously contributed their perspectives through face-to-face conversations, online interviews, review sessions, and written correspondence.[2]

List of participating organizations in this study:

- Nature Conservancy
- US Fish and Wildlife Service (Craig Brook Hatchery)
- National Tribal Council
- Department of Natural Resources of the Penobscot Nation, Indian Island
- Trout Unlimited, Kennebec Valley Chapter
- Trout Unlimited Maine Chapter
- The Kennebec Coalition
- Saco Watershed Collaborative
- Downeast Salmon Federation
- Natural Resources Council of Maine
- Wells Reserve at Laudholm

I acknowledge the Wabanaki tribes, particularly the Penobscot Nation, whose ancestral territory in the Penobscot River Valley forms the bedrock of this research. My overarching goal is to offer a nuanced portrayal of the context surrounding the onset of the dam removal movement in Maine and the two conflicting narratives it entails. In extending my appreciation to those who have enriched my understanding, I approach this chapter with the aspiration of honoring their viewpoints. As a French

[2] I express my gratitude to all the interviewees for their time and generosity, with a special mention to Thom Whiting, an angler and river activist, for the delightful addition of fresh Maine oysters during one of the field trips.

Canadian-based ally with a settler background, I endeavor to tread this narrative path with sensitivity and mindfulness, recognizing the weight of responsibility that accompanies recounting this story.

6.3 Case Studies

Since the 1950s, and even earlier for some, the decline of salmon sea-run populations has been unmistakable, confronted by persistent threats arising from both within and outside the watersheds. This chapter zooms in on the Kennebec and Penobscot Rivers as case studies, emblematic of the focal point of civic engagement aimed at revitalizing Atlantic salmon sea-runs.

These rivers illustrate both successful endeavors and ongoing restoration challenges. As the largest rivers in Maine, covering approximately one-quarter of the state each (see Fig. 6.1), they share a common history dating back to the glacial retreat 14,000 years ago. Home to the Wabanaki,[3] Maliseet, and Micmac (Mi'maq')[4] nations for the past 10,000 years, these rivers boast rich biodiversity across land, water, and air. Originally vital for timber, pulp industries, and power generation with over 100 dams on each mainstem, they have now transformed into popular destinations for rafting and sport fishing. These spaces provide recreational havens for families within and outside the cities they traverse.

Between 1983 and the early 1990s, a noticeable shift occurred from favoring pro-dam decisions to a more widespread acceptance of fish river habitat restoration. In 1992, the state imposed stricter regulations on anglers, limiting them to one Atlantic salmon per season. That year marked the initiation of environmental impact assessments and more rigorous conservation requirements for dam operators. The culmination of these efforts came in 1993 when the FERC denied Bangor Hydro's application to rebuild the Basin Mills naturally breached dam on the Penobscot River. This victory was a cause for celebration for the civic Penobscot River Coalition, particularly its members, spearheaded by the Veazie Salmon Club, who had actively collaborated and advocated for the dam's removal in the preceding years.

[3] The Penobscot and Passamaquoddy are federally recognized tribes.

[4] The Micmac tribe of the Sandy River, situated northwest of the Kennebec River valley, lacks recognition from the US Federal government. In contrast, the Micmac tribe of Aroostook County in the northeast of Maine, near the USA-Canadian border with New Brunswick, has obtained federal recognition. The substantial impact of this distinction is evident in the acknowledgment or lack thereof of their rights. Similar struggles are faced by other tribes, such as those in Vermont (Rolde, 2004; Taylor, 2020).

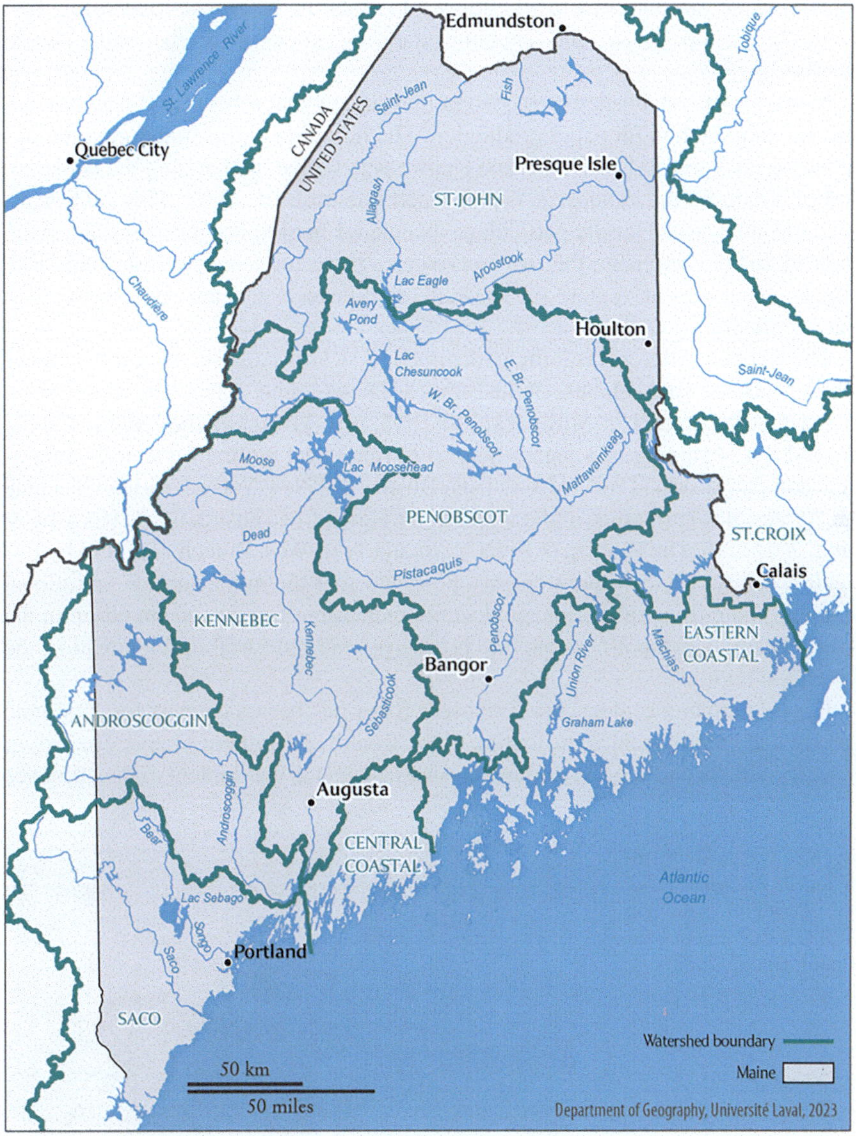

Fig. 6.1 Watersheds of Maine, showcasing the interconnected river networks across the state

6.4 The Lumbering Legacy of Northern Maine and Its Effects on River Ecosystems

Since the Industrial Revolution, Maine's rivers have been perceived as reservoirs of opportunities, serving as access routes for floating timber destined for paper mills and sawmills, harnessing hydroelectric power, facilitating the passage of ships constructed at the Bath Iron Works shipyard (along the Kennebec River) (Fig. 6.2) to reach the open sea, and enabling trade of goods with the wider world (Caldwell, 1983).

Maine's well-established river system played a pivotal role in its ascent as a significant national hub for paper production. The initiation of this transformation can be traced back to 1834 when the first charter was signed, authorizing the construction of a dam on the Penobscot West Branch (Hempstead, 1930). This dam facilitated elevated water levels, particularly beneficial for log driving along the river. Prior to this development, the river served as a route for canoeing from Bangor to Mount Katahdin—a region of spiritual significance for the Penobscot First Nation (Idem).

The Kennebec River was initially dammed in 1837 in Augusta, the capital city of Maine, with the establishment of the Edwards Dam (Crane, 2009). The construction of the first paper mill in Millinocket in 1898, also known as the "Magic City",[5] marked the beginning of a legacy shaped by the Great Northern Paper Company. This company, formed by investors from Boston and New York, eventually became the "largest newsprint mill in the world at the time of its construction" (Hornsby & Judd, 2015: 50). The Penobscot River, particularly its West Branch with four hydroelectric dams, provided the necessary power to fuel the mill's success and global renown. The mill, with its strategic location benefiting from a 100-foot drop in the river, became a symbol of economic prosperity and industrial development in the Downeast region.

For more than a century, the Penobscot River has been a catalyst for economic growth and employment, contributing to the wealth of Northern Maine. The city of Bangor, with its enduring Victorian mansions, stands as a testament to the affluence

Fig. 6.2 The author by the Kennebec River in Waterfront Park, nestled in the historic district of Bath. (Credit: Gravel, 2022)

[5]The Magic City Morning Star newspaper, embodying Millinocket's moniker, the "Magic City" (Woodbury, 2005), transitioned to a digital-only format in 2002, serving as an ironic testament to the paperless era in a region where paper production was central to many cities.

generated by the paper industry (Reilly, 2009). Despite changes in ownership and unsuccessful attempts to revive production, the closure of the mill had a profound impact on Millinocket and the region. Former mill workers recall a period where individuals could earn as much as they desired, reflecting the economic prosperity that characterized the city. The success of the Great Northern Paper Company not only provided income for the local population, including lumbering camp workers, but also played a significant role in sustaining the Maine logging industry, benefiting around 30,000 people each winter (Schmitt, 2015: 41).

Even after the forests within a 50-mile radius of Bangor were depleted, a phenomenon documented by Thoreau during his 1853 journey to Northern Maine (Schmitt, 2015: 42), lumbering persisted northward, with Bangor remaining the primary shipping point for wooden boards destined for global markets. The zenith of the lumber trade occurred in 1872, as timber was floated down the Penobscot River to supply over 400 sawmills in Old Town, Veazie, Orono, and Milford, catering to the global demand for wooden construction materials, even reaching as far as China (Ibid.). In its early days, with transportation options limited, the motto "let water do the work" emerged as a practical solution (Nelson, 1949: 113). Initially, oxen provided the motive force to convey logs from the woods, utilizing the ample snow cover during Maine winters to slide loads to waterways. To navigate tributaries and overcome river obstacles, Penobscot Indians[6] applied their river expertise to assist in log drives (Schmitt, 2015). While river logging officially persisted until 1972, by 1960, the majority of Maine pulpwood was being transported by trucks on the state's highways, a mode deemed more flexible and cost-effective (Hornsby & Judd, 2015: 69).

The environmental consequences of forest clear-cutting were recognized as early as 1869, identified as a contributing factor to the substantial reduction in river flows and the increase in water temperatures in the Bangor area (Eves, 1992; Hempstead, 1930). The construction of dams and waterlogging had profound effects on river health, with the decomposition of wood negatively influencing water quality parameters and oxygen levels.

The St. Regis Paper Mill in Bucksport, Maine, which operated from 1930 to 2014, played a pivotal role in shaping the town's identity as a company town situated along the Penobscot River. Its strategic location near the Gulf of Maine facilitated the convenient importation of essential inputs for papermaking, initially sulfur and later oil for paper drying, while also enabling global exports of finished products. During World War II, Time Inc. acquired the mill to secure a reliable paper supply for publishing its magazines in Boston and New York (Abbott, in Smith Ranzoni, 2017: 15–17).

The St. Regis Paper Mill, a significant supplier to printing companies in major cities, fueled the economic prosperity of families, led to the construction of homes, and employed up to 1200 individuals at its peak. The mill also established a bustling

[6] The term "Indians" is still actively used by members of Maine tribes and is officially recognized by the state government.

port, attracting ships and visitors worldwide, contributing to the town's vitality. In addition to shipping paper, the mill served as an export hub for Maine potatoes in the 1930s, reaching ports along the Atlantic coast from its steamship wharf. The closure of the paper mill, therefore, had a profound impact on the town, causing a ripple effect as it was not only the primary employer but also a major contributor to local taxes. In the anthology dedicated to the paper mill, editor Patricia Smith Ranzoni (2017) expresses boundless gratitude for the mill's historical significance and its role in shaping the community.

Smith Ranzoni's narrative is notably devoid of any discussion regarding workers' health hazards,[7] environmental impacts, or adverse effects of paper mill operations on Bucksport residents. Any inconveniences arising from the mill, such as noise and truck traffic, are portrayed positively, almost becoming a comforting cadence, a rhythmic heartbeat for the townspeople (Smith Ranzoni, 2017). The mill, in this context, becomes the standard for measuring various aspects of life, even geographical distances in the region are gauged in relation to it.

However, for newcomers entering the town of Bucksport in 2019, the initial impression didn't evoke scenic beauty but rather revealed the enduring marks of industrial activity that lined the entire waterfront of the Penobscot River. This industrial legacy manifested as a stark absence of trees and natural vegetation, replaced by expansive areas of asphalt, barracks, and towering chimney stacks. At the time of my visit, there were considerations for repurposing projects for the existing infrastructures.

Regarding the selection of the paper mill site on Indian Point, subsequently ironically dubbed Salmon Point, Smith Ranzoni (2017: 7, 9) recounts the presence of an Indian mound that obstructed the mill's construction and consequently had to be leveled. This mound was one of the few camping grounds of the "Red Paint People," ancestors of the Wabanaki tribes, especially the Penobscot Nation who lives along the Penobscot river (Idem). In an editor's note, acknowledgment is made that the land is part of the Wabanaki territory, extending from the Atlantic coast to the Appalachian range in the state of Maine. However, no argument is presented for the preservation or restoration of the site, contrasting with other historical sites like the Whaleback Shell Midden (Fig. 6.3a, b), located about 65 miles south along the coast of Maine on the Damariscotta River. The shell middens are the result of a millennium of Wabanaki tribe shellfish dining, and as the shells accumulated, they formed mounds upon which vegetation, including tall trees, grew. At their core, these mounds measure over 30 feet in height (Coastal Rivers Conservation Trust, website; Hardy, 2009).

The establishment of the paper mill led to the obliteration of the camping grounds belonging to the "Red Paint People," employing a "burnt land" strategy that

[7] In 2019, the local newspaper *Bucksport Enterprise* (May 9) reported that those who have developed cancer from working at the paper mill in an asbestos-filled environment will be able to get compensations since a law firm had started defending workers' interests. It is stated that the owner of the company did not know about the danger of asbestos. The period considered for claims is limited to the years prior to 1981.

Fig. 6.3 (**a** and **b**) Middens on the Damariscotta River. (Credit: Gravel, 2022)

irrevocably transformed the entire landscape. The disruption induced by logging activities, encompassing the building of roads, log driving, logging camps, and deforested regions, transformed hunting and fishing grounds into wastelands. It repurposed ancient trees into pulpwood and altered sacred territories into desolate areas. In his work "The Maine Woods," chronicling his three journeys on foot, portaging, and paddling through Northern Maine in 1846, 1853, and 1857, Henry David Thoreau expressed his sorrow at the sight of flooded forests and lifeless timber—both standing and adrift on lakes. Thoreau recognized the profound loss of environmental integrity. By 1923, all major lakes in Northern Maine had succumbed to flooding, rendering their shores inaccessible by boat due to the substantial driftwood accumulation, as observed by Thoreau's posthumous editor.

The environmental repercussions of paper mills' operations were far-reaching, affecting both water and inland areas. The byproducts, ranging from raw sewage from the workers' community to the waste chemicals discharged by the mills—comprising mercury, dioxin, and other hazardous substances—contributed significantly to environmental degradation. Additionally, the accumulation of sawdust and woodchips on riverbeds, the incessant movement of ships in the bay, the erection of over 100 dams exclusively on each river and its tributaries, and the widespread flooding of lands and islands collectively inflicted continuous damage on both terrestrial and aquatic ecosystems. This unrelenting assault was emblematic of the extractivist nature of the pulp and paper industry (Kurlansky, 2020; Acosta, 2011).

Industries associated with papermaking, like Holtra Chem in Orrington near the St. Regis Mill on the Penobscot River, played a significant role. Holtra Chem, operating from 1967 to 2000, produced bleaching agents for papermaking, utilizing mercury as an electrical conductor (Schmitt, 2015: 113). It disposed of mercury-laden sludges in unlined landfills along the riverbanks and discharged 16 pounds of mercury annually into the water during its operational years (Ibid.). Situated just 12 miles away from the St. Regis paper mill, its environmental impact extended beyond immediate proximity. Studies on bald eagles within the 124-mile watershed revealed extensive contamination diffusion (DeSorbo et al., 2018: 1516).

Closed in 2000, the Holtra Chem site, which significantly contaminated the river, estuary, and surrounding landscape, is slated for repurposing. New investors seek permits to establish an industrial zone on the eastern shore of the Penobscot River at the former site. A Bangor Daily News report in May 2019 highlighted a 10-year delay proposed for the plant facility's soil decontamination. The enduring contaminant burden and its ramifications on the watershed's flora and fauna remain inadequately assessed.

6.5 The Evolution of Energy Production and Demand in the State of Maine

Post-World War II, the average personal energy consumption of American citizens surged by 468% from 1946 to 1968 (Hornsby & Judd, 2015: 112). The rise in industrial development and urbanization subsequently triggered an increased demand for hydropower. Beginning in 1909, the Fernald law prohibited the state of Maine from selling its power beyond its borders (Craig in Smith Ranzoni, 2017). Senator Fernald, foreseeing the risk of Maine becoming a national power plant, mandated that all hydropower generated must meet an industrial energy demand within the state. Consequently, the St. Regis paper mill was strategically established to harness the hydropower from the Wyman Dam in Bingham, along the Kennebec River in northern Maine. The mill commenced operations upon the completion of power lines (Idem). After several unsuccessful attempts to repeal the law, it was finally repealed in 1955, a time when growing concerns about the state's environment led to reservations against further dam construction (Idem).

While most major hydropower installations were established by the mid-twentieth century, the completion of the Harris Dam on Indian Pond in the Upper Kennebec River in 1955 marked a significant addition. Acknowledged as the largest hydroelectric dam in Maine at that time, subsequent developments in alternative energy sources contributed to a diminishing emphasis on further hydropower expansion (Maine Memory Network website). Harris Dam elevated the water level of Indian Pond by 20 feet, expanding the surface area from 2 to 5½ square miles. The enlarged Indian Pond, involving the clearance and flooding of approximately 2500 acres of trees and bush, featured campsites accessible to the public and was promoted as a recreational and sport fishing destination (Idem).

6.6 Shifting Perspectives: Transitioning from a Utilitarian Worldview to Harmonious Coexistence with the Natural Environment

As early as the late nineteenth century, the declining salmon sea-runs in the Penobscot River prompted the establishment of the Craig Brook Hatchery in 1871, dedicated to producing salmon eggs for propagation and stocking efforts (Schmitt,

2015: 85, 110). Timid conservation initiatives in the 1920s and 1930s were followed by a state commission's report in 1945, commissioned by the Governor of Maine, emphasizing the severe impact of contaminating sources along the Penobscot River, including paper mills, residential sewage, and power plants that could harm the fish passing through turbines. The report recommended decontaminating the waters before constructing fish ladders: "the sport fishery was suffering due to the great scarcity of salmon" (Idem: 87).

By 1945, after the construction of the Mattaseunk Dam in 1939 on the Penobscot River, fish sightings were nearly nonexistent, despite the presence of a fish passage: "almost no fish had been seen" (Idem: 86). Commercial fisheries suffered, with salmon accessing only 2% of their spawning habitat in the watershed, leading to their closure in 1947 (Idem: 87). Rachel Carson (1951), the renowned marine biologist, writer, and conservationist, advised housewives to alter their cooking habits and choose fish other than salmon. The newly formed Atlantic Sea-Run Salmon Commission highlighted the salmon's plea to remove dams obstructing their access to spawning grounds upstream in the watersheds.

> The streams with the deepest gradient, the largest volume, and the most constant summer flow were best both for salmon and for industry...on the Penobscot River the salmon must pass over six dams—at Bangor, Veazie, Great Works, Old Town, Howland and Mattaseunk—in order to get access to the great spawning and nursery areas of the East Branch and the Seboeis and Wassataquoik tributaries (Salmon commissioners George Rounsefell and Lyndon Bond, 1947, in Schmitt, 2015: 87).

The construction of each new dam not only resulted in the loss of lands and fishing grounds for the Penobscot Nation but also had profound environmental consequences. For instance, the building of the Milford Dam and its neighboring dam on the Stillwater branch in 1905 led to the flooding of a 235-acre area, including islands and essential fishing grounds where Penobscot members had caught silver salmon for millennia (Idem).

By the 1950s, the Penobscot River had transformed into a substantial waste disposal area for mill waste, comprising cellulose liquor, wood, sawdust, and chemicals, a situation described as "long accepted." In 1929, members of the committee of paper industry representatives acknowledged this acceptance, stating, "It has long been accepted – and still is today – that our streams and rivers must serve as natural channels for the carrying off and removal of liquid waste materials" (Idem: 101).

The organic waste discharged into the river had a detrimental impact on river life, particularly affecting the oxygen levels crucial for aquatic fauna like cold-water salmon. Reduced oxygen levels are known to contribute to the decline in fish populations, leading to their demise. Additionally, downstream of the mill's operations, low acidic levels were identified. Despite this understanding, the industry vigorously advocated for its interests and often found support from state politicians who viewed industrial development as a driving force for regional growth and high employment rates.

The enactment of the federal Clean Water Act and the establishment of the Environmental Protection Agency (EPA) in 1972 provided the means and funding to address water pollution. This facilitated the construction of residential wastewater treatment facilities in cities and towns situated along the Penobscot and Kennebec Rivers, such as Bangor and Augusta (Courtemanch et al., 2023). Paper-making companies, compelled by regulations, were also obliged to invest in decontamination and prevention measures, exemplified by the Great Northern Paper Co., which invested $36 million in the 1970s to establish a wastewater treatment system (Schmitt, 2015: 118, 119). Subsequently, river pollution witnessed an 80% reduction, leading to the encouraging return of salmon. This positive development revitalized interest among sport fishermen, resulting in the proliferation of salmon clubs in the region and the attraction of new members from as far as Massachusetts, Connecticut, and New York.

6.7 River Conservation Advocacy and Discourse

By mid-century, a growing conservation awareness swept across the nation, led by figures such as artists and scientists, including Rachel Carson. The establishment of national parks reflected a shift in people's appreciation of nature, valuing it for its landscapes and recreational opportunities rather than solely for its natural resources. In the 1940s, the US Fish and Wildlife Service and the Atlantic Sea-Run Commission initiated a model for river restoration (involving federal and state agencies) with the aim of creating a haven for sport and potentially commercial fishing. The leaders of these services envisioned public awareness as a key solution to pollution problems. Starting in 1958, the construction of fishways became a requirement in all new dam projects, compelling builders to give equal consideration to wildlife as to other water resource development. Congress passed the Endangered Species Preservation Act in 1966, without listing the Atlantic salmon at that time (Schmitt, 2015).

The short-nose sturgeon, similar in habits to Atlantic salmon as an anadromous species, was listed earlier. However, Atlantic salmon faced opposition to its listing until 1995. In 1967, the Penobscot River earned the designation of a "Model New England Salmon Restoration River," although it was never included in the US network of Wild and Scenic Rivers. A significant legal battle in 1968, where the Department of the Interior sued the Federal Energy Regulatory Commission over proposed dams in Idaho, led to a landmark Supreme Court decision expanding the definition of the public good to include environmental values (Idem). The Clean Water Act of 1972 was enacted to sustain ongoing efforts to restore fish habitat and enhance water quality in rivers nationwide (Idem). The enforcement of water quality regulations from the 1970s onward played a crucial role in the gradual return of sea-run fish species, sending a strong message about the federal government's commitment to supporting river wildlife.

6.8 Restoring the Kennebec River: The Tale of the Initial Dam Demolition

The endeavors of Maine's river restoration advocates came to fruition with the removal of the Edward's Dam on the Kennebec River in 1999, marking the first dismantling of an operational hydropower dam in the United States (Lohan, 2019; Crane, 2009).

In the mid-1980s, the development of a novel water quality science bolstered initiatives advocating for river habitat restoration, encouraging residents in Maine to actively participate in water quality improvement (Courtemanch et al., 2023). Biologists and environmentalists from the Maine Natural Resources Council and Trout Unlimited, including the Kennebec Valley Chapter and the Maine Council of Trout Unlimited, began generating documents outlining the conditions and procedures for dam removal. This initiative received support from the University of Maine's Extension Office.[8]

Knowing that the FERC relicensing process of the Edwards Dam was scheduled for 1993, a Public Policy Working Group was established to examine dam removal. In the initial year, reports were drafted concerning the fisheries of the Kennebec watershed (Crane, 2009). Subsequently, during the second year, the topic of dam removal was deliberated through a series of public forums. The catalyst for envisioning the removal of dams, rather than solely constructing fish passages, arose from a reflection by a member of the Kennebec Valley Chapter of Trout Unlimited: "I wonder what the Kennebec would be like if Edwards Dam was removed" one of them contemplated (Brooke, personal communication, 2022).

KRAC was promptly established as a nonprofit organization, with the acronym representing the Kennebec River Anglers' Coalition, aiming to "put a crack in the Edwards Dam." (Idem) Given that numerous KRAC members were affiliated with the Natural Resources Council of Maine, a partnership was forged to initiate plans for dam removal. Recognizing the need for increased collective strength, additional conservationist groups joined forces, leading to the creation of an "unincorporated collaborative coalition" with the Kennebec Valley Chapter and the Maine Council of Trout Unlimited, American Rivers, and the Atlantic Salmon Federation Maine's Office "pooling resources and effort to promote dam removal at the head of tide, Edwards Dam," 40 miles from the estuary (Idem).

For the subsequent decade, KRAC successfully raised funds and benefited from the services of pro-bono lawyers to "support the legal case against Edwards Dam." (Idem) The Coalition actively participated in the relicensing process, presenting arguments to FERC that highlighted how the dam's negative impact on the Kennebec ecosystem outweighed the benefits derived from the power it generated (Idem; Crane, 2009) A compelling point was made: the dam produced only "3.5 megawatts of power, less than 0.1 percent of Maine's electricity" (Lohan, 2019). Furthermore,

[8] Steve Brooke, Project coordinator for the Kennebec Coalition and officer for Trout Unlimited Kennebec Valley Chapter, personal communication, May 2022.

it employed a small workforce, was aging, and posed safety concerns (Idem). The dam owners were also unwilling since the 1980s to engage in the Kennebec River fisheries restoration program and invest in fishways to facilitate the passage of shad and alewives around the dams (Crane, 2009). Under the pressure, the latter proposed to increase on-site power generation to 18 megawatts and invest in fishways (Idem). The technology of fishways has proven ineffective in the past, hindering downward migration and creating lethal obstacles for eels attempting to reach the ocean, as they would be inadvertently shred to pieces by the turbines (Fig. 6.4).

Ultimately, after submitting a comprehensive 7000-page document, the Kennebec Coalition garnered support from influential politicians, including Governor Angus King, who recognized dam removal as an opportunity for Maine to thrive both economically and ecologically. One prerequisite though for Congressional approval was to enhance hydroelectricity production at other locations along the river to ensure the project's power neutrality (Idem).

After numerous heated debates in public fora, considering the arguments for and against dam removal (see Table 6.2) and the perspectives of various stakeholders,[9] particularly riverine owners, the balance tipped in favor of river restoration. In a conclusive response, FERC mandated the owner of the Edwards Dam to relinquish the federal license and proceed with the dam's removal, a groundbreaking decision. The dam owner and FERC reached an agreement, leading to the removal of the dam in the summer of 1999 (Idem).

The dam removal movement that commenced in Maine and spread across the nation[10] marked a turning point, reshaping interests and the regional economic foundation. Since then, there has been a shift toward prioritizing jobs that have minimal

Fig. 6.4 Image of the KRAC logo
Description: The image above is the sewn patch that KRAC members received upon signing up (shared by S. Brooke, 2022)

[9] For a complete list of stakeholders and account of the story of the removal of Edwards Dam, see Crane (2009).

[10] The approval for the removal of the Elwha River Dam in Washington State was granted through successful federal legislation in 1992, but the actual dismantling did not occur until 2012 (Crane, 2009; American Rivers website). The primary motivation behind this decision was the restoration of salmon runs.

environmental impact over those reliant on resource extraction. A new era commenced, characterized by the paradigm of the river's rebirth and the comprehensive restoration of its habitat. The perception of nature underwent a shift. Initially regarded solely as a resource for extraction, it is now increasingly acknowledged as an inherent human and ecological habitat. This perspective promotes a coexistence that nurtures an evolving partnership, where the interests of both humanity and nature are not only protected but also actively supported.

6.9 The Penobscot River Coalition Project

Buoyed by the Kennebec River initial triumph, sport fishermen joined forces and allied with conservationist groups and the Penobscot First Nation to emphasize that prioritizing Penobscot River salmon was essential for public policy. The significance of salmon fishing clubs in the Bangor vicinity, with four major clubs within a 5-mile radius, should not be understated. Boasting over 1000 members, these clubs draw anglers from as far as Massachusetts. They constitute a pivotal stakeholder group, making a notable impact through their fundraising initiatives and their influential role in shaping discourse and persuading others (Butler & Taylor, 1992). Despite significant opposing interests from major companies and municipalities, supported by attorneys who attempted to instill fear among communities along the Penobscot River regarding potential costs associated with protecting the salmon habitat, it proved challenging to reverse the momentum generated by robust federal policies.

Since the Atlantic salmon was added to the Endangered Species list in 2009, the Federal government has mandated Environmental Impact Assessments for the construction of new dams and the relicensing of existing ones. The subsequent removal of dams in Maine primarily targeted the Penobscot River, where the Great Works (2012) and Veazie Dams (2013), both situated near Indian Island, were dismantled. Similarly, the Howland Dam, located upstream of the Penobscot Nation, was decommissioned without full removal. Instead, a bypass channel[11] was implemented to facilitate fish passage, thereby opening access to the Piscataquis River, a significant tributary (Courtemanch, 2022, personal communication; Taylor, 2020).

Following the completion of the Penobscot River Restoration Project, the Penobscot Nation experienced the immediate joy of witnessing the positive impacts on the river's configuration. Discovering the resurgence of fish, such as Shad near Old Town, allowed the Nation to revive ancestral fishing and canoeing practices

[11] As local approval was a prerequisite for Congressional support of the Penobscot River Restoration Project, the Coalition had to devise a resolution concerning the Howland Dam. The town of Howland, with a modest population of 1,200, resisted the dam removal but eventually voted to support the removal of the old tannery and the decommissioning of the dam. To mitigate any harm to young fish during downstream migration, the turbines were dismantled. Additionally, a fish passage was constructed to provide an alternative route around the dam (Taylor, 2020).

(Natural Resources Council of Maine, website; Taylor 2020: 127). "The ancestors are smiling today," exclaimed Butch Phillips, a respected Penobscot tribal elder and river ambassador, as he concluded the ceremony on 11 June 2012, with a prayer following the breaching of the Great Works Dam (Idem: 96).

The Penobscot River watershed serves as the residence for the ancestral and contemporary Wabanaki Penobscot Nation (see Fig. 6.5). The governing body of the Nation is situated on Indian Island, one among the 200 islands situated within the river basin. The traditional customs and practices of the Penobscot Nation members have been profoundly influenced by the industrial development that commenced in the 1850s in Maine, as their connection with the river intricately ties them to the territory (Hornsby & Judd, 2015). Their way of life is intimately linked to the river—they live by it, derive sustenance from it, and are inherently connected to its essence (Taylor, 2020).

A representative of the Penobscot Nation expressed the demands of Maine Tribal nations during a public hearing opposing the construction of the Basin Mills Dam (Schmitt, 2015: 155):

> We want returned to us sufficient runs of salmon, shad, and alewife to be able to harvest them in our waters. We are not anti-development, but in this case, the potential harm to our fisheries outweighs the benefit of this project to us and those people of Maine who inhabit

Fig. 6.5 Wabanaki riverine cultural landscape along the Penobscot river on Indian Island, showcasing the enduring connection of the Wabanaki people with the river. (Credit: Gravel, 2020)

the mid and upper areas of the Penobscot Valley who should be able to take for consumption those fish that we want restored.

In 2020, Peter Taylor contributed a historical and culturally inclusive account of the Penobscot River restoration. This comprehensive narrative captures the viewpoints of the coalition of organizations that worked together to acquire and dismantle mainstem river dams. Importantly, Taylor recognizes the pivotal contribution of the Penobscot Nation, ensuring that their perspectives and agency were central to the restoration process. The impact of indigenous worldviews on shaping the pro-restoration narrative and arguments has also been apparent.

While the Penobscot Nation did not assume a leadership role in the river restoration movement, anglers and conservationists spearheaded opposition against proposed dams slated for construction in the 1980s and 1990s on the Penobscot River. However, the Penobscot Nation played an early role in establishing the Penobscot River Restoration Trust in the early 2000s, collaborating with five conservation organizations: American Rivers, Atlantic Salmon Federation, Maine Audubon, Natural Resources Council of Maine, and Trout Unlimited (Taylor, 2020). The Penobscot River Restoration Trust was subsequently tasked with owning and removing dams. The board of directors of the Trust included representatives from the Penobscot Nation and all five organizations. The Nature Conservancy of Maine joined in 2006 (Idem). The collaborative efforts of the Penobscot River Coalition have been extensively covered by the news media for over two decades. This extensive documentation has played a pivotal role in shaping a pro-dam discourse, significantly impacting decision-making regarding the future of dams in Maine, as noted by Quiring (2020).

6.10 Indigenous River Advocacy Amid Tense Territorial Rights Dynamics

The 1992 publication "Penobscot River's Renaissance," edited by members of the Penobscot Salmon Club, recognizes the historical presence of Wabanaki Indians along the Penobscot River dating back 7000 years. However, it provides only limited acknowledgment of their contemporary existence and the ongoing legal debates surrounding their fishing sustenance rights. This exclusion is not innocuous; rather, it underscores the significance attributed to historical conflicts concerning fishing rights and the perspectives of settlers on natural resources and fisheries.

Until the 1980s, the Wabanaki tribes of Maine lacked official recognition, rendering them largely invisible—a situation that some Indigenous tribes, as seen in Vermont, continue to challenge (Rolde, 2004). Post the 1980s land settlement treaty in Maine, the Penobscot Nation received land and financial support to establish a reservation, designating Indian Island as the hub for tribal facilities and services. This land had been inhabited by their people for over 7,000 years, during which they engaged in hunting and fishing within its waters (Interview with J. Banks, 2019).

This area, along with 200 islands in the Penobscot River, was acknowledged as ancestral territory through the 1812 Treaty between Indigenous tribes and the state of Massachusetts—predating the existence of the state of Maine. The entire river ecosystem, encompassing humans, land and aquatic fauna, flora, insects, air, and water, is regarded within the Penobscot's worldview as a holistic entity. According to their perspective, conservation efforts should focus on protecting the entirety of the ecosystem rather than singular components, such as the Atlantic salmon (Interview with J. Banks). The belief is that restoration efforts benefit not only a specific species but contribute to the well-being of the entire ecosystem and various other species (Idem; Taylor, 2020) (Fig. 6.6).

In their advocacy for the river, the Penobscot Nation aimed to reaffirm their connection to the waters, extending their historical efforts in land reclamation. Technically owning the riverbeds and possessing sustenance fishing rights within the boundaries of their reservation—the entire Penobscot watershed—since the Law on Indian land claim settlement of 1980, this ownership has been contested in court since 1995 (Taylor, 2020).

The Penobscot Nation sought compensation from the state of Maine for the loss of river access and fishing rights. This loss incurred significant costs, as it compelled the community to forsake its traditional lifestyle, heavily dependent on river fish consumption. The absence of migratory fish returning to the river not only jeopardized the exercise of their sustenance rights but also posed a threat to the survival of their cultural traits and the possibility of intergenerational passing of traditions. Simultaneously, water contaminants made it impossible to regularly consume freshwater nonmigratory fish (Taylor, 2020).

Back in 1912, traditional salmon spearfishing techniques were banned in Maine, a response triggered by complaints from the Veazie Salmon Club located 12 miles downstream. Members of the club argued that the Native Americans were depleting

Fig. 6.6 A holistic representation of a salmon river habitat. (Source: Original painting by Juliette Richard, 2023)

salmon stocks, putting them at risk for sport fishing (Schmitt, 2015). Subsequently, fishing quotas were established as Atlantic salmon populations were recognized to be in jeopardy, and during the 1950s, it became intermittently illegal to catch any salmon from the river. Biologists reported no Atlantic salmon presence in the river between 1955 and 1966 (Idem). After the enactment of the Clean Water Act in 1972, when some salmon returned, it was determined that one salmon per family per year could be caught from the Penobscot River. The legislative confusion sparked competition among salmon fishermen, primarily anglers and netters, leading to tragic incidents where some Indian tribe members were killed while attempting to fish in disputed waters (Schmitt, 2015; interview with J. Banks, 2019). Tensions between the rest of the state and the reservation escalated due to fishing restrictions that were not adhered to by the Indians, leading to the criminalization of Wabanaki's fishing and hunting practices (Hornsby & Judd, 2015). Indian sustenance fishing rights, recognized in a 1980 treaty, were delineated within the boundaries of their reservation.

Together with the Penobscot River Coalition, comprising NGOs like Natural Resources Council of Maine, Maine Audubon, Sierra Club, Conservation Law Foundation, American Rivers, Sportsman's Alliance of Maine, and Trout Unlimited, the Penobscot Nation actively opposed the proposed construction of a new dam at Basin Mills in 1989. This project was eventually abandoned in 1998, following its rejection by FERC after a negative Environmental Impact Assessment. According to John Banks, Wabanakis engaged in negotiations with two essential conditions: first, the consideration of removing main stem dams, and second, the inclusion of their "friends" to avoid succumbing to divisive politics (Interview with J. Banks, May 2019).

The Penobscot Coalition evolved into a knowledgeable and science-driven network of activists, supported by federal agencies, dedicated to the restoration of native fisheries in Maine's largest watershed. Taylor's book (2020) chronicles the narrative of how they secured public support and raised funds, amounting to a $63 million project, to acquire and demolish dams. Convincing Congress necessitated demonstrating popular support within Maine, showcased through public event attendance, small donations, forums with river ambassadors, and engaging with local sentiments, fears, relationships, and love for the river.

Scientific reports on water quality and the river's condition formed the bedrock arguments favoring river restoration, facilitating the establishment of ecological corridors from the Katahdin head source to the Gulf of Maine. Enthusiasm for river restoration extended beyond freshwater fishermen to include lobster fishermen, emphasizing the interconnection between the river and the ocean. They highlighted the benefits to lobster fisheries with the resurgence of alewives and other sea-run fish in healthier rivers. After two centuries of disconnection, the Gulf of Maine stood to thrive once more. The media played a crucial role in disseminating these narratives of trust-building within Maine society, contributing to the widespread adoption of the discourse surrounding the river's rebirth.

6.11 Water Science Shaping the Discourse of River Restoration

The interview with John Banks, the director of the Department of Natural Resources of the Penobscot Nation, focused on the initiation of the Nation's water program and commenced with the query: How did the public in Maine come to accept dam removal? This question forms the crux of our examination of discursive narratives, perceptions, and acceptability. It delves into the societal transformation from an industrial economy to a postindustrial economy.

The discourse surrounding the restoration of the Kennebec River revolved around the notion that the value of a thriving river surpassed its exploitation (Lohan, 2019). The shift occurred as the public good ceased to be just linked to paper mills and became associated with a clean free-flowing river hosting salmon and 11 other fish species that have returned since restoration. This includes alewives, serving as a food source for various species. The river now provides amenities for citizens, including river trails, sport fishing, waterfront activities, scenic views,[12] white river rafting on Kennebec stretches, the resurgence of predatory birds like bald eagles and ospreys, and the scenic beauty of sunsets and dawns over clean waters. Local artist Kdb Dominguez[13] has played a significant role in documenting the evolution and showcasing the beauty of the Kennebec River estuary by capturing daily photographs.

An integral part of the pro-restoration narrative revolved around the scientific findings generated in the two final decades of the twentieth century. These results served as persuasive arguments presented to various groups to influence lawmakers and leaders in fish habitat restoration. Among these scientists, Daniel Kusnierz, who presented a talk at the 2019 Maine Sustainability and Water Conference titled "The Healing of Rivers: Water Quality over the Last 25 Years," played a crucial role. Employed by the National Tribal Council, Kusnierz has been associated with the Penobscot Nation Water Resources Program since the early 1990s. During an additional meeting at the Natural Resources Department on Indian Island in April 2019, in the company of the former director of this facility, John Banks, Kusnierz highlighted the grim state of affairs during the era of paper mills. Back then, ten paper mills along the Penobscot River main stem were responsible for river foam, unpleasant odors, and the release of dioxin, which accumulated in fish tissues in the Penobscot River. The years 2001, 2004, and 2007 marked significant occurrences of algal blooms, attributed to cyanobacteria, with mills identified as sources of phosphorus. In his capacity as a water scientist, he conducted research alongside his

[12] The city of Augusta's initial neglect of the river during construction remains a poignant memory for river activist Brian Graber, now with American Rivers (Lohan, 2019: 5).

[13] Bath artist Kdb Dominguez has spent the last decade photographing Merrymeeting Bay and the Kennebec River estuary as part of her Wilderland Climate Change Project. This project provides a visual narrative depicting the transformation of aquatic habitats through a combination of painting and photography. https://www.kdbwilderland.com/

team and set up monitoring platforms at a paper mill in Dolby Pond. Following the cessation of mill operations in 2008, no further blooms were observed. However, concerns lingered regarding phosphorus (P) stored in sediments, often released during dam removals and subsequently flushed by the current, climate change, and nutrients. He noted that the Maine Department of Environmental Protection (DEP) had yet to finalize a nutrient criterion.

According to him, the Penobscot Nation has been unable to exercise its fishing sustenance rights due to ongoing water pollution, accompanied by health warnings. The tribe, taking charge of its own consumption rates for mercury, dioxins, and PCBs, has devised an "Eating Wild Food Safety Series" (2017). This informational brochure elucidates the traditional diet and provides advice to community members on safe consumption levels for various fish species, such as brook trout or landlocked salmon (Reese, 2022; Eiring, 2022). There is a glimmer of hope for the resumption of traditional fish consumption from the river, as dioxin levels in fish have reportedly decreased since 2003, as per Kusnierz (personal communication, 2019).

Kusnierz's research involves biomonitoring studies, utilizing aquatic insects and benthic fauna as indicators of water quality, given their exposure to bioaccumulation. After the removal of the Veazie Dam in 2013 on the Penobscot River, there was a noticeable rise in oxygen levels in the water and a decrease in water temperature, demonstrating positive changes in all parameters that support river ecological restoration.

Table 6.1 below compares the impacts of dam removal on water parameters at Great Works and Veazie sites, both situated downstream of Indian Island, between Old Town and Bangor (where (+) indicates an increase and (-) denotes no change). Notably, the removal of Great Works, located upstream of the Veazie site, did not independently result in improved water oxygen; the removal of the Veazie Dam was crucial to enhancing water quality downstream of Indian Island.

Drawing on a decade of water quality analysis, Kusnierz asserts that the water now complies with new standards. During this period, there was an improvement in water classifications across various rivers in Maine. Notably, the West Branch Penobscot upgraded from D to C to B, the East Branch Penobscot improved from B to AA, the mainstem Penobscot advanced from C to B, and several tributaries moved

Table 6.1 Comparative analysis of water parameter changes following dam removal at Veazie and Great Works

Parameters	Great Works 2012	Veazie 2013
Oxygen	(−)	+
Temperature	(−)	(−)
Aquatic insects	+	+
Daily T range	+	+

Source: Kusnierz (2019)

from B to A, with some reaching AA.[14] Regarding salmon, Kusnierz notes a positive shift from the absence of salmon to their presence: he concludes "there was none, and now there is some." Despite the efforts made to bring back salmon runs, other mainstem river dams continue blocking their migration, and many agree that "problems are also somewhere else, in the Northern Atlantic".[15]

Another preoccupation I had at the time of the interview was whether members of the indigenous community participated in the water analysis. He answered: "Tribal members are involved in all steps of the program, and Wabanaki students at U. of Maine come in as summer interns and move up in water analysis/wastewater treatment. Most staff of the program are tribal." (Reese's blog, 2022) The collaboration between the Penobscot Nation's Department of Natural Resources Water Resources Program and the University of Maine has played a pivotal role in fostering cross-cultural understanding, establishing a shared vocabulary, and recognizing the significance of indigenous environmental science in ecological restoration (Quiring, 2020). This decolonizing approach empowers Penobscot Nation's members to interpret and use data, enhancing their understanding of their reality and articulating their perspectives informed by water analysis criteria. The Penobscot River Restoration Program stands out as one of the pioneering initiatives in the United States with a central focus on socio-environmental justice on a watershed scale (Idem).

The pro-restoration narrative found resonance with the perspectives and conservationist values held by outfitters and fishing clubs, particularly exemplified by the Veazie Salmon Club. Instrumental in initiating the 'Stop Basin Mills' movement in 1989, club members engaged in fundraising activities to oppose the construction of the Basin Mills Dam on the Penobscot River (Schmitt, 2015: 154). Drawing from their experience battling Bangor Hydro against the reconstruction of the Bangor Dam on the Penobscot River after its 1977 breach, anglers played a pivotal role in shaping the Penobscot Coalition established in 1988.

Recognizing that the reconstruction of Basin Mills[16] would jeopardize the fate of Atlantic salmon in the Penobscot River, Coalition members were motivated by the findings of studies conducted by the US Fish and Wildlife Service. These studies indicated that under the conditions of the 1980s, salmon had a 70% chance of recovery. However, if another dam were constructed between the Veazie and Great Works Dams, those chances would plummet to below 40% (Idem: 153). In the pre-1990s era, a fish activist remarked that the "public is unaware that there is opposition to Basin Mills" and that "they tend to accept the dam as inevitable" (Ibid.).

As outlined by Schmitt (2015), while numerous organizations and biologists focused their efforts on opposing the Edwards Dam on the Kennebec, the Penobscot

[14] It is to be noted that AA waters are pristine waters while D waters are a nuisance. The 1985 updated classification is based on the river carrying capacity and the results for each parameter at a given time. Maine's actual water classification system is inspired by the research work of David Courtemanch, a Nature Conservancy officer who agreed to give me an interview in March 2019.

[15] Interview with Mark McCollough from the Fish and Wildlife Service at the Craig's Hatchery in Orland, March 2019, supported by Kurlansky, 2020.

[16] The Basin Mills Dam was a structurally compromised construction that steadily deteriorated, eventually breaching under the force of the current in 1910 (Butler & Taylor, 1999).

Coalition successfully enlisted various partners, including the Sierra Club, the Conservation Law Foundation, and American Rivers, the latter of which had identified the Penobscot River as the third most endangered on their 1989 list. The Sportsman's Alliance of Maine and the Natural Resources Council of Maine, each boasting approximately 9000 members, eventually joined the cause, followed by Trout Unlimited (Idem).

Beyond the water quality criteria that formed the core of the dam removal issue, fish activists recognized the river's carrying capacity—indicating the pressure a river could sustain in terms of pollution and obstacles to water movement. They leveraged the ecosystem argument when persuading the people of Maine, using catch phrases like "There's only so much a river can take" to encapsulate their perspective. Additionally, some influential leaders, including Jerry Marancik, the Maine Fishery Program coordinator for the US Fish and Wildlife Service, invoked the Rights of Nature argument, stating:

> We're restoring the salmon for the sake of the fact that they should be allowed to survive. The salmon should exist here, the same as the peregrine falcon, or the bald eagle or the whooping crane [...] If there is some angling effort or enjoyment that some can get from that, that's fine, but that's not why we're doing it (Butler & Taylor, 1992: 49).

The ideology of rights of Nature posits that all living, nonhuman entities should possess the right to persist without facing threats from human activities. Many countries have embraced the rights of Nature, drawing inspiration from indigenous nations worldwide (Further discussion on this topic will be provided later in this chapter) (Vega Cárdenas & Turp, 2021; ACLRC website).

In response to critics attempting to downplay the restoration efforts as solely driven by fishermen's interests, it became crucial for key figures like Marancik to emphasize that the ecological argument was the primary motivation. While anglers offered moral and occasional financial support, most of the funding originated from the federal government, primarily through the US Fish and Wildlife Service (Butler and Taylor, 1992).

As the environmental conservation paradigm gained strength in the 1990s, arguments supporting fish habitat restoration became increasingly prevalent, capturing the attention and support of journalists and the people of Maine. While fish activists still contend with arguments centered around power generation and water level control, there is growing recognition that the overall benefits of dam removal for the river and air ecosystem outweigh these concerns. In the current landscape with more sustainable and renewable energy sources available, the small contribution of these dams can be effectively replaced by solar, wind, or geothermal power generation. This is the current battleground for the Downeast Salmon Federation, which is actively challenging the Toronto-based company Brookfield, the owner of two dams obstructing fish migration on the Union River in the east—another river in Maine offering prime salmon habitats after the Penobscot and Kennebec rivers.

Brookfield Renewables owns a total of 38 dams across the state of Maine, with four of them on the Kennebec River, between Waterville and Skowhegan (Fig. 6.7), were targeted by the state of Maine for demolition (Carpenter, 2024; Woodard, 2021).

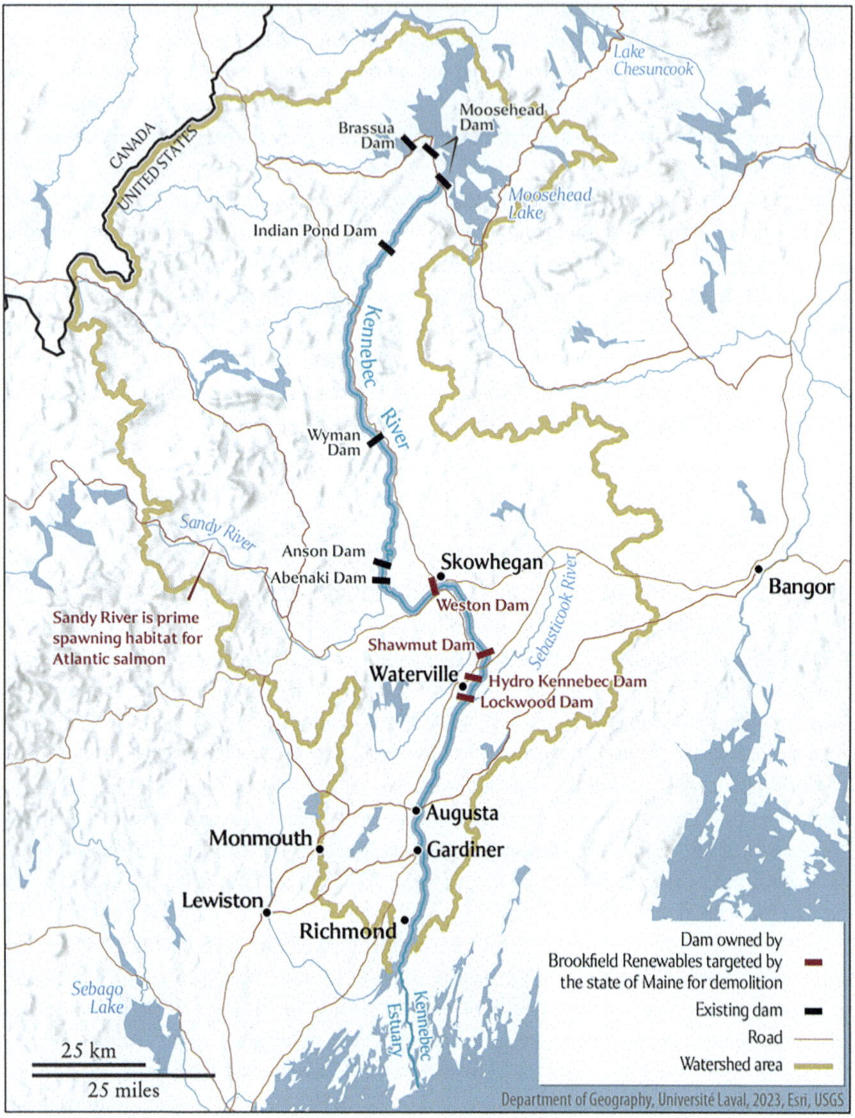

Fig. 6.7 The Kennebec River with designated dams slated for demolition. (Source: Inspired by Woodard, 2021)

In response, Brookfield filed a lawsuit against the state of Maine in 2021, accusing them of misconduct and interference in the relicensing process by the Department of Marine Resources—a strategic move by the company to delay and discredit the FERC examination process, especially as its dams' licenses are expiring and require FERC approval for continued operation (Idem). Governor Janet Mills, representing the state of Maine, first dismissed the lawsuit as meritless and urged Brookfield to adopt a more responsible role in environmental protection before reversing her decision (Carpenter, 2024). Brookfield, a subsidiary of the $600 billion global asset company Brookfield Asset Management, contends that the planned fish lifts at these dams will sufficiently protect Atlantic salmon and other species, enabling them to reach spawning grounds (Idem). However, the state and environmental groups argue that dam removal would enhance the prospects of restoring salmon runs, emphasizing that dams continue to pose a threat to fish year-round. This concern is exemplified by the need for Maine Department of Marine Resources workers to rescue stranded fish in the pools below the dams, as illustrated by the incident at Lockwood Dam in Waterville in June 2021 (Idem) (Fig. 6.8). In total, Brookfield produces about 87% of Maine's hydroelectric power and 21% of its wind energy (Carpenter, 2024).

Many small dams have already been removed from tributaries and river mainstems, serving as sources of inspiration for future restoration efforts. A notable recent example is the removal of the China Lake Outlet Stream Dam, connected to the Kennebec River via the Seven-Mile Stream. After a hiatus of 239 years, during which Maine was still part of Massachusetts and the Revolutionary War had just ended in 1783, alewives and other anadromous fish species have returned to China Lake "under their own steam" (Munroe-Hultman, 2022).

Fig. 6.8 View of Lockwood Dam on the Kennebec River in Waterville. (Credit: Gravel, 2022)

Landis Hudson, the executive director of Maine Rivers, a nonprofit organization dedicated to conserving the state's waterways, expressed, "Alewives were knocking on the door" (Idem). Maine Rivers took on the coordination and direction of what evolved into the China Lake Alewife Restoration Initiative. Hudson emphasized, "We had to go for it because the fish were waiting" (Idem). The community had long been preparing for the return of alewives, not salmon, as the China Lake Outlet Stream is located above the Waterville Lockwood Dam, making realistic goals a priority. We will delve into more details later.

A recent triumph celebrated by the Downeast Salmon Federation and its collaborators involves the removal of the defunct Meddybemps Powerhouse, a 75-year-old dam that previously served as a barrier between the Dennys River and Meddybemps Lake. The elimination of this dam is poised to contribute to the restoration of the lake's fisheries, particularly benefiting alewives. The town of Meddybemps stands to gain economic advantages, as alewives are in demand as lobster bait.

Notably, the land formerly occupied by the powerhouse has been returned to the Passamaquoddy tribe, allowing its members to reconnect with their ancestors' territory. In 2021, the tribal historic preservation officer expressed the significance of this event, stating that a ceremony had taken place, and they planned to pay homage to their surroundings. The successful realization of this project underscores the collaborative efforts involving the federal and state governments, the Passamaquoddy tribe, and other partners, including descendants of the dam builder (DSF website; Fox 22 Bangor news) (Table 6.2).

Table 6.2 Arguments for and against dam removal in the state of Maine

Arguments for Dam removal	Arguments against Dam removal
Value of natural amenities is greater than value of power generation (12 species could benefit and whole habitat restoration) Hydroelectric power production with numerous small dams on rivers is damaging for ecosystems The return of the Atlantic Salmon Opportunities for Indigenous tribes to resume their traditional fish consumption and lifestyle Recreational and commercial fishing revival possibilities Citizens' reappropriation of rivers and enjoyment of waterfront activities and nature (trails, canoe, fishing, rafting) Increased value of waterfront property	Loss of jobs with utilities closure Riparian property damages: erosion, flooding and low water levels may occur Loss of recreational activities on dammed lakes Loss of value for riparian properties Hydroelectric power is needed for growth Love and fondness of man-made infrastructures and associated scenery (dams seen as a symbol of progress and modernity)

Source: N. Gravel (2019)

6.12 Arguments in Favor of Dams

In the contentious debate over potentially reconstructing Basin Mills on the Penobscot River between 1989 and 1993, the Friends of Basin Mills rallied substantial support, gathering around 3000 advocates who staunchly believed in the positive outcomes associated with controlling rivers. Their argument rested on championing the right to "safe, clean, low-cost electricity" (Schmitt, 2015).

Dams, historically constructed for electricity generation, flood control, and water level management, have multifaceted perceived benefits. Riparian property owners, particularly those with residences along riverbanks, often resist dam removal due to concerns about potential adverse effects on their property from variable water levels.

Support for dams extends from individuals practicing nautical sports to those who appreciate the scenic beauty created by waterfalls resulting from the presence of dams, fostering a sense of connection to the surrounding landscape. Some express affection for hydroelectric facilities and associated industries, such as paper mills. While some man-made historical structures hold cultural value for their owners and communities, the costs associated with maintenance, repairs, and legal compliance for fish migration may outweigh considerations for their preservation. A case in point is the Fort Halifax Dam, upstream of the removed Edwards Dam on the lower Kennebec River. In 2008, the dam's owner deemed building fishways cost-prohibitive, leading to the decision to dismantle the dam, thus connecting the Sebasticook River with the Kennebec. Some members of the community vigorously fought this decision by the company, which delayed its removal by a few months (Munroe-Hultman, 2022: 4). The Sebasticook River now supports the largest run of river herring on the east coast of the United States, exceeding 6 million fish (Natural Resources Council of Maine, website; Crane, 2009).

6.13 A Balanced Approach to Fish Habitat Restoration

The strategic measures adopted by fish advocates since the 1990s to encourage non-hydroelectric dam owners to adopt responsible practices provide valuable insights into the decision-making dynamics within a river governance framework. A comprehensive approach involving community outreach, fish ecology education, advocacy, and consensus-building plays a pivotal role in garnering support from communities situated along lakes and rivers for the successful implementation of fisheries restoration projects.

Fish ecologists and activists organize public hearings to elucidate the fish's needs, outline the benefits for the community, plan for river restoration and fish recolonization (assisted or natural), and assess with geomorphologists the release of sediments and flow in the river channel and how fish habitats could be impacted (Allen et al., 2016). In their Condit Dam removal study in the State of Washington, Allen et al. (2016) warn that planning for anadromous fish recovery prior to dam

Fig. 6.9 A fish ladder at Nequasset Dam in Woolwich. (Credit: Gravel, 2022)
Description: Displaying design flaws and functional inadequacies, this inefficient fish ladder, constructed in 2014 by engineers without consultation with local fishermen, replaced the one built in 1955. It raises concerns since then among local fishermen and environmental advocates. Notably, Woolwich residents mandated a fishway through the dam during a town meeting as early as 1760. The dam, situated on Nequasset Brook, ensures a consistent water level for Nequasset Lake, serving as the primary water source for the city of Bath

removal is crucial. This may necessitate relocating fish to upstream river stems and lakes to ensure spawning occurs post-dam removal, as the violent flow of sediments rushing through the breach can negatively impact spawning grounds.[17] Alternatives to dam removal, such as the construction of intelligent fishways, are also presented. Unlike many existing fishways that are dysfunctional and fail to mimic the natural shapes and dynamics of the river, intelligent fishways are designed to be recognized and utilized by fish.[18] An interview with an experienced alewife fisherman at Nequasset Dam in the Kennebec River watershed highlighted the lack of consultation with local stakeholders during the construction of a fish ladder. The ladder, depicted in Fig. 6.9, has proven ineffective for fish migration due to its design at an

[17] A National Geographic video illustrates the sediments rush after breaching a dam https://www.youtube.com/watch?v=4LxMHmw3Z-U&ab_channel=NationalGeographic

[18] Maine public work employees are active in culvert replacement at road crossings to enhance habitat connectivity for fish and wildlife. Stream Smart culverts—structures designed to be sufficiently large (more than 1.2 times the width of the stream) and mimic a natural stream bed in depth, slope, and material—ensure unimpeded fish migration throughout the year, even during varying water levels (2019 Water conference, Augusta).

angle that fails to replicate natural river curves, resulting in fish getting stuck in the corners.

Among the responsible options available to dam owners supporting fish run restoration are fish ladders and lifts (functioning as elevators), man-made channels dug around dams, and the complete removal of the dam to facilitate fish and eel access to their spawning grounds. It's important to highlight the challenge faced by eels on their downstream migration from a lake, particularly due to operating turbines. Determining the precise timing of their migration is challenging, making it difficult to shut off turbines in time to prevent their harm, even with the best cooperation from dam owners. Fish lifts can serve as a lifesaving alternative for eels when deemed feasible (Natural Resources Council of Maine, website).

Communities find economic benefits in the return of fish, with opportunities for sport fishing, commercial fishing (alewives), recreational activities, nature observation, and tourism, helping them stand out and attract visitors (Idem). The resurgence of fish breathes life into streams, ponds, and lakes, fostering the presence of new predatory birds (ospreys, herons, bald eagles) and other fish species (American shad, bass, sturgeon, and eels). Despite the initial focus on reviving Atlantic salmon, as suggested by many interviewees, even if the plan doesn't unfold as expected, the result is often a diverse array of fish species and a flourishing environment (Idem). Considering these positive outcomes, one might question the arguments that could potentially oppose a river renaissance.

Following extensive collaboration with communities along the China Lake Outlet Stream, the team dedicated to fish restoration faced a pivotal negotiation with the town of Vassalboro, the owner of the dam serving as the final obstacle before alewives could access China Lake. The suspense and pressure on the fish advocates' team were palpable after years of persuading other towns to support fish habitat restoration.

To address the concerns of the town's residents, a consulting committee was established to deliberate on the future of the dam at the outlet of China Lake. The committee carefully considered residents' perspectives, leading to the decision to construct a Denil fishway[19] at the dam. This intermediate solution allowed alewives to migrate upstream while ensuring that residents could continue enjoying the lake's aesthetic and recreational values (Munroe-Hultman, 2022: 5). For decades, residents had actively participated in swimming and boating activities on China Lake, rendering the removal of the dam a nonnegotiable option for them to avoid lowering the water levels (Idem: 6). However, it is acknowledged that while fishways have their merits, they fall short of the outcomes achieved through dam removal, constituting a "second best to complete restoration" (Idem: 6).

The program leader, Landis Hudson, employs a "coffee-table approach," emphasizing open discussion once granted permission to assess a site for restoration. To

[19] The Denil fishway, with its smaller ecological footprint compared to other fish ladders, not only facilitates fish migration but also became a local attraction, allowing onlookers to observe the fish during their journey.

elucidate, consider the following illustration. The fish passage engineer systematically identifies all available options, presenting conceptual designs to stakeholders in a reciprocal conversation. This dialogue involves receiving concerns and providing feedback, all with the overarching goal of eliminating obstacles to fish passage. The program director terms this method an "iterative process" or "creative compromise," highlighting its significance in river restoration. According to the director of Nature Conservancy, this approach is deemed "key to restoring a river," emphasizing that there is no universal solution for stream connectivity. She states, "Aquatic connectivity doesn't always mean dam removal. You have to alleviate the anxiety of landowners and municipalities" (Molly Payne Wynne, quoted in Idem: 5).

6.14 The Future of Atlantic Salmon River Restoration in Maine

One lingering question persists: How much additional effort will be required to reclaim the remaining mainstems and significant spawning areas of Maine's salmon rivers? In 2015, President Obama implemented a rule aimed at water quality protection, equipping the EPA with tools to regulate agricultural activities, industrial contaminants, and lawn pesticides flowing into rivers. Unfortunately, this rule was overturned by President Trump in 2019, leading to a decline in federal support in the following years (Friedman & Davenport, 2019).

However, the Biden administration has introduced landmark legislation, signed in 2021 and 2022, focused on addressing national priorities with a particular emphasis on improving infrastructure, addressing climate concerns, promoting racial equity, and bolstering the economy (Idem). In 2022, the National Fish Passage Program is slated to support 40 projects in 23 states and Puerto Rico, emphasizing partnerships aimed at eliminating barriers to fish passage and restoring interconnected rivers and coasts. This initiative includes funding for the removal of two dams on the Sabattus River, a tributary to the Androscoggin, within the Kennebec River watershed (Idem).

Facilitating access to upstream lakes like Sebasticook, Sabattus, and China, along with critical tributaries such as the Sandy River, is imperative for the restoration of Atlantic Salmon and the other 11 sea-run fish species. This holistic watershed approach, emphasized by Courtemanch in a personal communication in 2022, is crucial for unlocking the full potential of nursery areas, with the Sandy River alone identified as presenting half of the potential spawning grounds for the Kennebec River below Madison Dam, as reported by the Atlantic Sea-Run Salmon Commission in 1986 (Ibid.). Maine public opinion has undergone an unprecedented shift toward environmentalism, but certain opposing narratives and reluctance to alter human practices have impeded the movement to entirely restore Atlantic salmon sea-runs, preventing them to reach their spawning grounds. The privileging of individual rights over the common good remains present in riparian communities, and ongoing efforts to convince and educate their members are essential. Despite historical progress, much work remains to be done.

What about salmon's rights? In the 1980s, some fish advocates argued for fish habitat restoration primarily for the well-being of salmon. In the contemporary context, there is a growing movement that recognizes the Rights of Nature, providing a platform where the rights of salmon could be duly represented.

Can the Rights of Nature perspective effectively contribute to environmental protection, and will the American legal system accommodate it? The response to the plight of dying salmon could potentially mark a crucial juncture for the Rights of Nature movement, as suggested by Frank Bibeau, an attorney for the White Earth Ojibwe, as quoted by Lee (2022). Salmon, as a species, requires clean water and access to spawning grounds for survival.

In another part of the country, the Sauk-Suattle tribe initiated a lawsuit in 2021 at a tribal court and won in 2023 against the city of Seattle during a relicensing process with FERC (Idem; Surma, 2023). This legal action was pertinent as the 30-year license for three dams without fish passages, owned by Seattle City Light Co. on the Skagit River, is set to expire in 2025. According to Bibeau, utilizing rights of nature lawsuits could serve as a mechanism to challenge the traditional judicial system and recalibrate power dynamics between states and tribal nations. Bibeau emphasizes, "Dams limit the flow of water, which reduces the flow of nutrients and other vital aspects of the river's ecosystem and [is] catastrophic for species like salmon, which rely on those areas to spawn" (Bibeau in Lee, 2022). He further contends, "Rights of Nature are an acknowledgment that everything is connected. Protecting salmon helps protect the rivers, the orcas, the people living near them [...] The case rests on the simple concept that salmon deserve the right to live and be defended" (Idem). In the 2023 settlement, the "City of Seattle agreed to provide passageways for the fish around hydroelectric dams on the Skagit River." The dams, built in the twentieth century, violated a treaty according to the Sauk-Suattle tribe that gave them the legal right to fish salmon in the Skagit River. It was dubbed the "rights of salmon" case because it sought a declaration by the city that the fish "possess inherent rights to exist, flourish, regenerate and evolve, as well as inherent rights to restoration, recovery and preservation" (Surma, 2023: 1).

6.15 Concluding Remarks

The double-sided narrative presented here recounts the historical transformation of Maine's two most important salmon rivers, the Kennebec and the Penobscot Rivers. Over the centuries, these rivers underwent a process of confinement and ecological disruption due to the proliferation of numerous dams catering to industrial needs. Discourse analysis sheds light on the clash between the established extractivist paradigm justifying these structures and the emerging environmentally conscious paradigm, characterized by growing coalitions of organizations. This ideological shift triggered conflicts over divergent viewpoints, values, and interests. Despite progress made to free these rivers, achieving complete connectivity to reach spawning grounds remains the next significant challenge to truly aid the Atlantic salmon. So much work has been done, only to be halted just a few miles from the finish line!

Supported by federal agencies, forward-thinking politicians, media, scientists, Indigenous nations, committed legal advocates, international organizations like the Atlantic Salmon Federation and the North Atlantic Salmon Conservation Organization (NASCO), and angler/outfitter clubs, the movement for free-flowing rivers navigated through constructive debates and public hearings. This collaborative endeavor laid the foundation for the acknowledgment of a new economic identity for Northern Maine. It recognized the vibrant and emerging sectors of sport fishing, fisheries, outdoor tourism, all centered around river amenities. Governor Angus King exemplified this shift by endorsing the removal of the Edwards Dam on the Kennebec River. McPhee (in Crane, 2009: 144) recounted the historical moment:

> Schools of striped bass had been seen by construction workers nosing up to Edwards Dam that morning. A big Atlantic salmon broached and—two feet out of water—just hung there, watching. A high-school bell was mounted on a post near the micro-phones, and Babbitt, King, and others—on schedule with the speeches and the downstream recession of the morning tide—began to ring it. And ring it. Evan Richert, of the Maine State Planning Office, shouted into the microphones, "Set the fishery free!" He did not say, "Set free the river!"

The deep attachment to fisheries by fish activists, Indigenous nations, and the broader population fueled a movement to salvage and propel those salmon rivers into a postindustrial era. The 1990s ushered in scientific and technological advancements, coupled with heightened ecological awareness, paving the way for envisioning a different world built on new pillars: emerging renewable and sustainable energy technologies, reduced demand for paper and lumber, a renewed appreciation for leisure and clean waters, and a profound rediscovery and embrace by the people of Maine of their rivers.

Multiuse trails, such as the Augusta Greenway (see Fig. 6.10) and the historically named "Papermill Trail" in Lisbon within the Androscoggin watershed, have emerged along Maine rivers after their revitalization. These trails not only contribute to public well-being by promoting a healthier lifestyle and rekindling the connection with rivers as integral components of "blue cities," but have also become focal points for renewed civic engagement. The Maine Trails Coalition, established in 2019 and operated by volunteers, is actively planning for active transportation, proposing trail corridors connecting existing paths along the Kennebec River to facilitate active commuting between Augusta and the coastal region of Brunswick (Maine Trails Coalition website).

A noteworthy outcome of the river restoration movement is the increased market value of riverfront properties. Augusta residents, who once turned away from the Kennebec River due to persistent sewage smells, are now embracing it as part of their daily lives. The return of ospreys and bald eagles, the latter formerly on the US endangered species list, symbolizes the positive transformations observed by the

Fig. 6.10 Segment of the Augusta Greenway, offering a picturesque view across from the city of Augusta. (Credit: Gravel, 2019)

community. The interconnectedness of watersheds and airsheds, as articulated by Indigenous tribes' cosmology, underscores the holistic impact of river restoration.

While many initiatives to restore crucial fish habitats were initially spearheaded by passionate Maine anglers, the success of these efforts hinged on broader public acceptance. Groups like the Kennebec River Valley Anglers' Coalition (KRAC) and the Penobscot Coalition, in collaboration with the Penobscot nation, played an instrumental role in fostering river mindfulness. However, the movement gained momentum with diverse forms of civic engagement, including contributions from university students, faculty, civic groups, artists, writers, and journalists. This collective engagement created a conducive backdrop for the emergence of the Maine free-flowing river movement.

Numerous governmental entities, both federal and international, are actively participating in the salmon restoration movement. Their engagement serves as a beacon for groups in other regions, inspiring the quest for a fresh equilibrium between the obsolete extraction-driven economic paradigm propelled by human desires and the burgeoning ecocentric ethos, beckoning humans to adopt a more modest role

Fig. 6.11 A sunset over the Kennebec River estuary, highlighting the delicate beauty of a habitat deserving conservation efforts. (Credit: Gravel, 2022)

alongside nonhuman living beings. It is a crucial reminder that rivers, too, are alive and deserving of our care and consideration. While the Atlantic salmon symbolized the pursuit of clean waters, success, according to many, depends on revitalizing the entire river ecosystem (see Fig. 6.11).

References

Acosta, A. (2011). *Extractivism and neoextractivism: Two sides of the same curse*. The Transnational Institute. Retrieved at: https://www.tni.org/files/download/beyonddevelopment_extractivism.pdf. Accessed 13 Jan 2024.

Allen, M. B., Engle, R. O., Zendt, J. S., Shrier, F. C., Wilson, J. T., & Connolly, P. J. (2016). Salmon and steelhead in the White Salmon River after the removal of Condit Dam–planning efforts and recolonization results. *Fisheries, 41*, 190–203. https://doi-org.acces.bibl.ulaval.ca/10.1080/03632415.2016.1150839. Accessed 26 May 2024.

Atlantic Sea-Run Salmon Commission. (1986). *Atlantic Salmon management in the Kennebec River: A status report and interim management plan*. 9 p. Retrieved at: https://www.maine.gov/sos/cec/rules/13/188/ken-salmon.pdf. Accessed 13 Jan 2024.

Butler, J. E., & Taylor, A. (1992). Penobscot River Renaissance. In *Restoring America's Premier Atlantic Salmon Fishery*. Down East Books.

Butler, J., E. & Taylor, A. (1999). *Penobscot River Renaissance: Restoring America's Premier Salmon Fisheries. Camden, Me, Silver Quill Press.*

Caldwell, B. (1983). Rivers of Fortune. In *Where maine tides and money flowed*. Guy Gannett Publishing.

Carpenter, M. (2024). Brookfield: Maine dam king. The Maine Monitor. Accessed August 31, 2024. https://themainemonitor.org/brookfield-maine-dam-king/

Carson, R. L. (1951). *The sea around us*. Oxford University Press.

Courtemanch, D. L., Davies, S. P., Johnson, E. S., Shaffner, R., & Suitor, D. (2023). Before and after the clean water act: How science, law, and public aspirations drove seven decades of progress in Maine water quality. *Maine Policy Review, 32*, 1. Retrieved at: https://digitalcommons.library.umaine.edu/mpr/vol32/iss1/4. Accessed 13 Jan 2024.

Crane, J. (2009). "Setting the river free": The removal of the Edwards Dam and the restoration of the Kennebec River. *Water History, 1*, 131–148. https://doi.org/10.1007/s12685-009-0007-2. Accessed 13 Jan 2024.

DeSorbo, C. R., Burgess, N. M., Todd, C. S., Evers, D. C., Bodaly, R. A., Massey, B. H., Mierzykowski, S. E., Persico, C. P., Gray, R. B., Hanson, W. E., Meattey, D. E., & Regan, K. J. (2018). Mercury concentrations in bald eagles across an impacted watershed in Maine. *USA Science of the Total Environment, 627*(2018), 1515–1527. Retrieved at: https://briwildlife.org/wp-content/cache/mendeley-file-cache/309090a5-7bf9-37f4-b517-34c18c4846f4.pdf. Accessed 13 Jan 2024.

Eiring, Z. (2022). U.S. Fish and Wildlife Service Northeast Region blog. In *Traditional knowledge of Penobscot Indian Nation influence on wildlife projects*. Retrieved at: https://usfwsnortheast.wordpress.com/2017/11/15/traditional-knowledge-of-penobscot-indian-nation-influence-on-wildlife-projects/. Accessed 13 Jan 2024.

Eves, J. H. (1992). 'Shrunk to a comparative rivulet': Deforestation, stream flow, and rural milling in 19th-century Maine. *Technology and Culture, 33*(1), 38–65. Retrieved at: https://doi.org/10.2307/3105808. Accessed 13 Jan 2024.

Friedman, L., & Davenport, C. (2019). Trump administration rolls back clean water protections. *The New York Times*. Retrieved at: https://www.nytimes.com/2019/09/12/climate/trump-administration-rolls-back-clean-water-protections.html. Accessed 13 Jan 2024.

Gravel, N. (2019, May 27–31). *"Let the river flow": Dams removal and the emergence of new forms of watershed governance in the State of Maine, U.S.* Paper presented at the Annual Conference of the Canadian Association of Geographers (CAG), Winnipeg.

Hardy, K. (2009). *Notes on a lost flute. A field guide to the Wabanaki*. Down East Books.

Hempstead, A. G. (1930). *The Penobscot Boom and the West Branch of the Penobscot River*. The University of Maine. Electronic Theses and Dissertations. 3290. Retrieved at: https://digitalcommons.library.umaine.edu/etd/3290/. Accessed 13 Jan 2024.

Hornsby, S. J., & Judd, R. W. (Eds.). (2015). *Historical Atlas of maine*. The University of Maine Press.

Kurlansky, M. (2020). *Salmon. A fish, the earth, and the history of their common fate*. Patagonia Books.

Kusnierz, D. (2019). *The healing of rivers: Water quality over the last 25 years*. Paper presented at the Maine Sustainability and Water Conference, Augusta.

Lee, D. (2020). *Restigouche: The Long Run of the Wild River. Fredericton, N.B. Goose Lane Ed.*

Lee, J. (2022). How rights of nature and rice could stop a pipeline. Grist. Accessed August 27, 2024. https://grist.org/indigenous/how-rights-of-nature-and-wild-rice-could-stop-a-pipeline/

Lohan, T. (2019). How Removing One Maine Dam 20 Years Ago Changed Everything. The Revelator. Accessed August 27, 2024. https://therevelator.org/edwards-dam-removal/#

Munroe-Hultman, L. (2022). *The little stream that could and will again*. U.S. Fish and Wildlife Service. Retrieved at: https://www.fws.gov/story/little-stream-could-and-will-again. Accessed 13 Jan 2024.

Nelson, F. A. (1949). *Lest We Forget*. Falmouth Publishing House.

Quiring, T. (2020). *Using engaged rhetorical methods to understand and inform collaborative decision making about dams and restoration in the Penobscot River Watershed*. University of Maine Electronic Theses and Dissertations. 3157. Retrieved at: https://digitalcommons.library.umaine.edu/etd/3157/. Accessed 13 Jan 2024.

Reese, D. (2022). *ESAL blog. Native Americans and other experts monitor the Penobscot Watershed in Maine*. Retrieved at: https://www.esal.us/penobscot-watershed-in-maine/. Accessed 13 Jan 2024.

Reilly, W. E. (2009). *Remembering Bangor. The Queen City before the Great Fire* (American Chronicles). History Press.

Rolde, N. (2004). *Unsettled Past, unsettled future. The story of Maine Indians*. Tilbury House.

Schmitt, C. (2015). *The President's Salmon. Restoring the King of Fish and Its Home Waters*. Down East Books.

Smith Ranzoni, P. (Ed.). (2017). *Still Mill. Poems, Stories and songs of making paper in Bucksport, Maine 1930–2014*. North Country Press.

Surma, K. (2023). Lawsuit asserting "the Rights of Salmon" ends in a settlement that Benefits the Fish. *Inside Climate News*. https://insideclimatenews.org/news/06052023/lawsuit-asserting-the-rights-of-salmon-ends-in-a-settlement-that-benefits-the-fish/. Accessed 26 May 2024.

Taylor, P. (2020). *From the Mountains to the Sea. The Historic Restoration of the Penobscot River*. Islandport Press.

The Bucksport Enterprise. (2019, May 9). *St. Regis Workers May [Can] Seek Asbestos Cash* (pp. 1 and 5. vol. 28, no 20). Inaccessible online.

Vega Cárdenas, Y., & Turp, D. (2021). *Une personnalité juridique pour le fleuve Saint-Laurent et les fleuves du monde*. Éditions JFD.

Woodard, C. (2021). Canadian Dam owners Sues Maine over alleged effort to remove 4 Kennebec River Dams. *Press Herald*. Retrieved at: https://www.pressherald.com/2021/09/28/canadian-dam-owner-sues-maine-over-alleged-effort-to-remove-4-kennebec-dams/. Accessed 13 Jan 2024.

Woodbury, D. A. (2005, August). *How Great Northern Paper Fell. Bangor Metro* (pp. 52–54). Retrieved at: https://woodburydavid.files.wordpress.com/2017/02/bm1.pdf. Accessed 13 Jan 2024.

Websites

Alberta Civil Liberties Research Center (ACLRC). *Rights of Nature*. https://www.aclrc.com/rights-of-nature. Accessed 13 Jan 2024.

American Rivers. (2014). Elwha River's Health Rebounding As Biggest Dam Removal in History Nears Completion. https://www.americanrivers.org/media-item/elwha-rivers-health-rebounding-biggest-dam-removal-history-nears-completion/ Accessed 13 Jan 2024.

Coastal Rivers Conservation Trust. https://www.coastalrivers.org/trail/whaleback-shell-midden-state-historic-site/. Accessed 13 Jan 2024.

Department of Environment and Science, Queensland. (2021). Denil fishway, Wetland *Info* website. https://wetlandinfo.des.qld.gov.au/wetlands/management/fish-passage/technologies/fishway-options/denil/. Accessed 13 Jan 2024.

Downeast Salmon Federation. (2023). https://www.mainesalmonrivers.org/. Accessed 13 Jan 2024.

International Rivers. *Movement for free-flowing rivers*. https://armecofront.net/en/news/movement-for-free-flowing-rivers-grows/. Accessed 13 Jan 2024.

Maine Memory Network. *Making electricity*. https://www.mainememory.net/sitebuilder/site/2134/slideshow/1261/display?format=list&prev_object_id=3535&prev_object=page. Accessed 13 Jan 2024.

Maine Trails Coalition website. https://mainetrailscoalition.org/about. Accessed 13 Jan 2024.

National Geographic, Condit Dam demolition in 2011 in Washington State https://www.youtube.com/watch?v=4LxMHmw3Z-U&ab_channel=NationalGeographic. Accessed 26 May 2024.

Natural Resources Council of Maine (NRCM). *Restoring a free-flowing Kennebec River*. https://www.nrcm.org/programs/waters/kennebec-restoration/. Accessed 13 Jan 2024.

NRCM Great Works and Veazie Dams Removed From Penobscot River https://www.nrcm.org/programs/waters/penobscot-river-restoration-project/great-works-dam-veazie-dam-removal/. Accessed 13 Jan 2024.

Penobscot Nation Wild Foods Safety Series Brochure. https://www.penobscotnation.org/images/natural-resources/Documents/PIN-Fish-Brochure-Final-Draft-01162018R.pdf. Accessed 13 Jan 2024.

Part II
The Role of Social Innovation and Public Policies in Designing Transition Landscapes

Chapter 7
Wild Native Bees at the Farm: The Evolution of Farming Practices (1800–2020) and Their Impacts on Pollinator Habitat in New England

Abstract This chapter delves into the historical transformations of New England homesteads from 1800 to the present, with a focus on how changing agricultural practices have shaped the rural landscape. Factors such as modern farm implements, industrialization, and forest regrowth have negatively altered crucial pollen reservoirs that supported bee populations. Using a landscape study approach, this qualitative assessment explores optimal practices in rural territory management to ensure the future of wild native pollinators. The pre-modern New England farmstead model emerges as a valuable resource for reimagining our relationship with agriculture and restoring ecological habitats for pollinators. Addressing the need for regional food sufficiency for both humans and bees involves cultivating diverse crops and implementing changes in farming practices that will resonate throughout the landscape. The chapter concludes by outlining steps for agroecological conversion—drawing on Cuban examples—, the revival of abandoned farms, and the reinstatement of food citizenship, underscoring the interconnectedness of human and nonhuman components.

Keywords Agricultural diversification · Agroecology/Agroecological conversion · Beescaping practices · Cuban organic farming · Ecosystem services · Farming practices · Food citizenship · Landscape study · New England · New England homesteads · Nonhumans · Pollen reservoirs · Pollinator population decline · Pollinators' habitats · Regional food sufficiency · Rural communities · Rural territory management · Native Bees Wild pollinators

7.1 Introduction[1]

In urban cultures, insects have often been marginalized, deemed either useless or, worse, destructive. Our societies have dedicated considerable energy and resources to combatting insects perceived as threats to our chosen lifestyles. Whether it is

[1] This chapter was first presented at the Rural Quadrennial 2019 Working Landscapes and Livable Communities Conference at the University of Vermont under the title: GRAVEL, N. (2019) Farmland habitat loss for native pollinators: What are the prospects for recovery? A landscape study of the Kennebec county, Maine, U.S

N. Gravel, *Tending Nature*, https://doi.org/10.1007/978-3-030-47062-3_7

termites nibbling at wooden house boards, wasps disrupting leisurely afternoons on a patio, or mosquitoes causing discomfort during outdoor activities, the instances are plentiful. Meanwhile, the contributions of other beneficial insects (of which 99% fall into this category) have either been overlooked or wrongly grouped with harmful pests, neglecting their crucial roles in ecosystems (Tallamy, 2018: 109). Consider creatures like spiders that help control unwanted bugs or earthworms that enhance soil fertility. Pesticide manufacturers have facilitated the widespread purchase of products aimed at efficiently eradicating all insects, driving them away from urban environments, golf courses, and agricultural fields. In certain cases, sterilizing chemicals are employed in commercial agriculture during the initial stages of land preparation to eliminate any potentially harmful organisms in the soil that could impede crop growth.

The postwar era saw the emergence of a campaign to eradicate “the enemy” from gardens, utilizing surplus chemicals developed during World War II. The subsequent “Green Revolution” of the 1960s, championed by large chemical companies, resulted in unintended consequences for the invertebrate population (Tompkins & Bird, 1989). The notorious neonicotinoid-based Roundup, employed as a systemic pesticide in coated seeds, posed significant threats to bees and various insects worldwide (Hunter & Lightner, 2016). The detrimental impact of pesticides on beneficial insects prompts reflection on why the link between pesticides and their impact on useful bugs was not grasped earlier. Beyond the disruptive presence of chemical products in the environment, other factors, including the lack of food, shelter habitat, disease, and access to quality water, contribute to the challenges faced by pollinators.

This research, conducted in three regions of Maine, USA (south, mid-coast, and northeast), underscores the importance of transitioning toward ecologically diverse smallholder farming. Embracing organic or natural farming practices, multipurposed and multicrop agroecosystems, is crucial for restoring in-field biodiversity and preserving the vital environmental service of pollination. Additionally, managing forests to establish a balanced relationship between wooded areas and cropland, in a fragmented patchwork pattern at the regional scale, is essential to foster the thriving of undergrowth vegetation and pollinator populations (Hunter & Lightner, 2016; Nalepa & Colla, 2023).

The forthcoming timeline delineates the evolutionary adaptations of New England homesteads spanning from the 1800s to the contemporary era, drawing insights from diverse historical sources such as the Maine Agricultural Society, Farmers’ Almanacs from the latter half of the nineteenth century, the 1930s Maine State Farms accounts, and accounts from anthropologists and historians chronicling the transformation of rural communities and agricultural practices since 1830. The chapter culminates in a presentation of essential steps for effecting agroecological conversion toward more sustainable agri-food systems, rejuvenating abandoned farms, and reinstating the food citizenship that characterized the pre-modernization farm era. The old New England farmstead emerges as a pivotal model for reevaluating our connection with agriculture and food within a local foodshed, contributing significantly to the restoration and reconnection of ecological habitats for pollinators. In essence, a region that sustains human life will also nourish bee populations. Achieving regional food sufficiency necessitates cultivating a diverse array of crops and instigating substantial changes in farm design, layout, and practices. Agroecologists converge on the consensus of reverting to a multicrop rotation agricultural system, including fallow periods.[2] The interconnections between these components will be explored in the ensuing discussion.

7.2 The Impact of Agricultural Modernization on New England Landscapes

Agriculture in New England has undergone continuous evolution, with the shift from ecologically diverse and functionally varied farmsteads to specialized and mechanized farms resulting in the loss of a diverse and fragmented landscape. Nineteenth-century farms, characterized by diversification and waste reduction practices, provided a range of products and services for homesteaders (Sanford, 2015). The advent of the Maine Central Railway between 1856 and 1862 marked the onset of modernization, facilitating easier trade with regional markets, including Boston (Falk, 2012). This transition fueled demands for fresh produce, cereals, and meat from nearby cities. The premodern rural landscape, marked by diversified small-scale agriculture, humanized spaces, and increased biodiversity, contrasts with today's countryside. A typical New England farmstead featured a vegetable garden, small orchard, pasture, and cropland often enclosed by stone walls near the main house. The remnants of these historical farm ecosystems, such as stone walls, old foundations, and preserved barns, provide valuable insights into past agricultural practices. Analyzing these historical landscapes becomes a source of inspiration for reimagining contemporary agriculture, emphasizing agrobiodiversity in rural areas. This chapter focuses exclusively on wild native pollinators, elucidating distinctions between non-native honeybees (Apis mellifera) and wild native bees. The research aims to propose a model for sustainable agriculture in a North American context, fostering the healing and recovery of farming ecosystems, with a primary emphasis on wild native pollinator populations and biodiversity (Sanford, 2015; Falk, 2012).

7.3 Data Collection and Analysis Methods

During my sabbatical year in 2018–2019, I had the opportunity to explore rural landscapes in Maine and Vermont extensively. These experiences formed the basis for my research, with reference sites primarily located in this part of New England.

[2]The decline in farmland biodiversity during the twentieth century in Canada and the USA was closely linked to the disappearance of fallow practices (Agriculture and Agri-Food Canada, 2014). In the pursuit of "productivity," chemical inputs were used to replace the need for soil regeneration, a decision that proved to be a mistake.

As a guest professor at the University of New England in Biddeford, southern Maine, I had the chance to delve into the region's unique characteristics. Additionally, visits to the University of Maine in Orono, a hub for bee specialists, enriched my understanding of local ecosystems. The University of Maine played a pivotal role in initiating a research network focused on understanding the factors contributing to honeybee decline and advocating for their conservation. A graduate student from the University of Maine introduced me to organizations such as the Xerces Society, US Bee Cities, and Bee Campuses initiatives, which are influential contributors to civil society, spearheading the ecological movement for bee conservation not only in New England but throughout the United States.

The data collection methods employed for this study involved a comprehensive and diverse dataset. These methods encompassed first-hand data collection through interviews with various informant groups, including academics, federal services staff, civil society organization personnel, and rural key informants across all three regions of Maine under investigation. In addition to interviews, the following data collection techniques were utilized:

- Review of Local and Scientific Literature.
- Comparative Analysis of Topographic Maps on a series spanning from 1898 to 2018, specifically focusing on Kennebec County in Central Maine.
- Field Visits and Observations to assess the present-day landscapes in relation to the historical data from the topographic maps. This approach facilitated a deeper comprehension of the various historical layers that constitute the current rural landscapes.
- Visits to historical sites, regional museums, and antique stores were conducted to gain valuable insights into rural life during various historical periods, with a specific emphasis on the period following 1820. These visits included the examination of old books and farmers' almanacs discovered within these antique stores.

These comprehensive data collection methods ensured the acquisition of a well-rounded dataset, enabling a thorough exploration of the rural history and evolution of the study regions in Maine.

7.4 Conceptual Framework: Agroecology, Agri-Food Systems, and Foodsheds

Agroecology represents a sustainable approach to food cultivation, drawing upon local knowledge and principles of ecological sustainability. It endeavors to foster the vitality of soils and the well-being of those engaged in agricultural practices, all while upholding respect for the environment. This approach is deeply rooted in ecological thinking and emerges from a comprehensive perspective on the agri-food system. As Griffon (2013: 31) elucidates, "Agroecology is both a mindset and a mode of action, serving as a philosophy of our connection with nature, and as a collection of agricultural production techniques that prioritize environmental preservation" (translation from French by the author).

Agri-food systems earn their name by necessitating a conscientious approach that encompasses all the activities necessary to bring food from field to fork (FAO website) including food production, distribution, consumption, and waste management, all while acknowledging the interdependence of various stakeholders within a specific region or at the global scale. A localized agri-food system is a cohesive system at the regional or local level, characterized by interconnectivity that serves as a foundation for collaboration and partnerships. When localized within a single bioregion, this integrated system is commonly referred to as a foodshed. To transition from unsustainable food production systems to environmentally sustainable ones, it becomes imperative to cultivate a shared sense of awareness, make choices that reflect sustainable principles, carefully manage ressources by reducing losses and wastes, and actively participate in the local food system as committed social actors.

When we rely on the "global foodshed" (Gliessman & Rosemeyer, 2010: 347), a practice that has become prevalent with the advent of globalization, we become detached from the origins of our food and the destinations of food waste. Many concur to say that to achieve sustainable agri-food systems, it is crucial to reconsider this globalized approach and reconnect with the locales where our food is grown and where food waste is managed (McGreevy et al., 2022).

A foodshed refers to the geographical area where our food is produced, transported, traded, consumed, and ultimately disposed of (Kloppenburg et al., 1996). The term draws an analogy to watershed systems that collect rainwater to nourish rivers. When individuals choose to consume from their local foodshed and acknowledge their role within the entire food system, not merely as end consumers, they are actively cultivating food citizenship (Food Ethics Council website). In a world where many societies have become reliant on imported food, adopting the habit of consuming from one's local foodshed may require adjustments and changes in culinary practices. We have grown accustomed to the convenience of purchasing everything we desire at the supermarket without considering the origins of our food.

What steps are required to initiate transformation and reclaim the lost biodiversity while creating ecologically sensitive food systems and practices that safeguard the territory and its inhabitants?

7.5 Considerations Regarding Scale and Territorial Management

Efficiently managing the distribution of forests, croplands, and fallow land finds its most effective implementation at a regional geographic scale, particularly within the framework of bioregions. This approach offers a pathway to restore the vitality and diversity of ecosystems, ultimately benefiting both the local fauna and flora, as well as the overall energy balance within the bioregion. Furthermore, the planning and execution of strategies for managing and controlling invasive plant species can be more effectively coordinated at this regional scale.

The term "bioregion," derived from "biological region" (Sale, 1996; Berg, 2001), represents a scale that encompasses not only the human population but also the broader biosphere. Within a bioregion, humans share resources, confront common biophysical challenges and barriers, and cultivate a shared culture and language. They reside in close proximity to one another and foster a sense of identification with the territory they inhabit. Ideally, efforts are marshaled to establish a self-contained regional system encompassing production, consumption, and waste management within these bioregional boundaries.

While the concept of a bioregion is not a recent development, its significance was underscored by the work of Peter Berg, who founded the Planet Drum Foundation in San Francisco, USA, during the 1970s. This era marked the emergence of debates surrounding sustainable development, necessitating a clear definition of operational terms (Celnik, 2016; Berg, 2001; Stuppia, 2022). The idea of "living within the boundaries of one's bioregion" thus came to signify the practice of meeting one's needs by utilizing the resources and offerings of one's own region without imposing negative externalities on neighboring regions.

In the context of studying the territory of Maine, bioregions could align with the state's eight tourist regions, a division that appears to enjoy consensus (also employed by the Maine Government and other mapmakers). This division effectively captures the topographical characteristics and shared lifestyles of its residents, making it a practical framework for exploring and implementing sustainable regional practices.

When those territorial definitions coincide with the legal operating framework of counties and states, the task seems easier; otherwise, cross-border cooperation (between two states, for example) and intraregional cooperation will become necessary. In the case of Maine, as there are more counties than regions, each region corresponds to more than one county, except for the Greater Portland and the Southern Maine regions (Fig. 7.1).

7.6 Challenges Related to Declining Pollinator Populations and Their Vulnerability

Since 2004, the world has grappled with a concerning issue known as the "bee survival problem," marked by a notable decline in honeybee colonies—a phenomenon often referred to as "bee colony collapse disorder" (FAO, 2016). This issue has received extensive media coverage, often featuring stark images of deceased honeybees near their hives. However, this emphasis on honeybees tends to oversimplify the intricate tapestry of pollinator populations. Honeybees are commonly depicted as valuable allies to farmers, using rented beehives frequently transported from one

Fig. 7.1 Map of Maine Touristic Regions

field to another to fulfill crop pollination needs. In contrast, wild native bees have rarely been the focal point of research or media attention, possibly due to their "invisible" presence in agriculture. Most of them are solitary, residing in the soil or cavities, unmanaged by humans, and not honey producers.

Nonetheless, it is crucial to recognize that certain wild native bees, such as bumblebees, play a vital role as highly efficient pollen carriers, serving as alternative and supplementary pollinators—particularly during times when honeybee populations are vulnerable (Idem). In fact, these wild native bees can be considered our "Plan B" should honeybees face extinction. While wild native bees may confront similar detrimental factors as honeybees, accurately assessing their populations proves challenging. The International Union for Conservation of Nature (IUCN) Red List identifies nearly 20% of native bumblebee species as being at risk of extinction (van Vierssen Trip et al., 2020). Their exposure to stressors like pesticides may be less frequent compared to honeybees, although exceptions exist. As an example, in the 1970s, the use of pesticides over blueberry plantations in the Canadian province of New Brunswick had a profound impact on bumblebee populations, requiring 3 years for their populations to recover from this bee decline (The Xerces Society, 2011: 79–80).

The practice of relocating beehives from one field to another is known to place significant stress on honeybee populations, disrupting their natural connection to the environment and their ability to navigate effectively. Stationary pollination services are encouraged on farms that are growing a mix of crops and fruit trees to protect honeybees' health (Hunter and Lightner, 2016: 118). While native and honeybees may seem like competitors in scenarios where pollen stocks are low, they actually serve complementary roles, ensuring the continuity of essential pollination services (Martins et al., 2015).

Estimates underscore the critical role of pollinated plants in human diets, encompassing grains, fruits, nuts, and vegetable crops. Remarkably, approximately 75% of all plants rely on animals for pollination (The Xerces Society, 2011: 15). An astounding 35% of the world's food crops are entirely reliant on animal pollinators to reproduce (USDA website). Notably, specific fruit production, including apples, strawberries, and blueberries, would face dire prospects without the essential role played by bee pollinators. In certain instances, the pollination crisis has become so acute that apple producers in the Indian Himalayas are considering converting to less bee-dependent crops (Luthra et al., 2023). Manual pollination has become the only viable option for others, as seen in pear and apple orchards in the Sichuan region of China (Melvin, 2022; The Xerces Society, 2011: 12). In both cases, pollinator populations have declined or disappeared from these areas, partly due to the introduction of European honeybees (Apis mellifera), which competed with native bees (Luthra et al., 2023). Despite their lesser adaptation to the local climate of the Himalayan foothills, European bees have been favored over domesticated native bees because of their greater honey production (Idem). Indian producers have reported that the use of pesticides to increase yields and compensate for labor shortages in orchards is also responsible for the decline in honeybee populations (Idem). With climate change pressures and decline in honeybees, orchard workers in the Sichuan region are compelled to ascend ladders and painstakingly pollinate flowers individually—a labor-intensive and costly process that underscores our profound dependence on these diminutive lifeforms. Innovations like bee drones have been developed for the same purpose, seeking to mitigate the impact of declining bee populations on pollination services. Overall, those orchard producers' livelihoods and food security are threatened, with a possible halving of the fruit production in the years to come.

During my visits to the University of Maine in 2019, it was disheartening to learn that the rusty patched bumblebee (Bombus affinis), once thriving in New England, was considered extinct in that region. The majority of sightings of this bumblebee species were concentrated in its native state of Minnesota, as evidenced by observations made at locations like the Minnesota Landscape Arboretum in 2021 (Minnesota Landscape Arboretum website), as well as in Illinois and Indiana. As part of the petition submitted in 2013 by Jepsen et al. from the Xerces Society for Invertebrate Conservation and the University of Minnesota to have the rusty patched bumblebee listed as an endangered species under the US Endangered Species Act, a historical map was presented. This map illustrated the distribution of the rusty patched bumblebee from 1884 to 1999 (Colla, 2016). The data revealed that this bee species was once widespread in New England and along the eastern seaboard until 1999, as well as in the provinces of Quebec and Ontario and across the central United States. However, despite intensified search efforts after the year 2000, only 13 sightings were recorded within the national territory (Jepsen et al., 2013).

A study conducted in 2011 came to a grim conclusion, stating that the "relative abundance of the rusty patched bumblebee has declined by 95%," a striking contrast to historical figures (Cameron et al., 2011, cited in Jepsen et al., 2013). This drastic decrease in population has been commonly attributed to a fungal disease, an explanation that gained support from the National Research Council in 2007. This disease was initially contracted in Europe during the early 1990s when hives were imported to the United States for managed pollination in tomato and blueberry fields. The transmission of this pathogen likely occurred from managed bumblebees to their wild counterparts, placing three bumblebee species at risk, with the fourth species, Franklin's bumblebee, having already gone extinct as a result (Ibid.; interviews with entomologist Francis Drummond, University of Maine, 2019, and Eric Venturini, then officer at the Xerces Society, 2019).

In addition to the fungal disease, several other known factors have likely contributed significantly to the decline of these bumblebee populations. These factors include habitat loss, urbanization, and the use of pesticides, with systemic pesticides being particularly detrimental to bees as they affect their nervous system. The combination of these factors has posed a severe threat to the survival of these bumblebee species.

Why should we care about the rusty patched bumblebee? Bumblebees in general are keystone species in many ecosystems that support humans and animals alike by being among the best pollinators for crops such as blueberries, cranberries, and tomatoes. By doing so, they allow wildflower reproduction, as well as fruit and seed production, which feeds different species, from songbirds to grizzly bears. They are therefore key species in the food chain, and we cannot afford to keep losing biodiversity!

The preservation of the rusty patched bumblebee is of paramount importance for several compelling reasons. Bumblebees, including the rusty patched bumblebee, play a vital role in various ecosystems and serve as keystone species. Here's why we should care about their survival:

- Crucial Pollinators: Bumblebees are among the most effective pollinators for a wide range of crops, including blueberries, cranberries, and tomatoes. Their pollination services are invaluable to agriculture, ensuring the production of these crops, which are essential components of our diets.
- Wildflower Reproduction: Bumblebees contribute significantly to the reproduction of wildflowers. Their role in pollinating native plants is vital for the maintenance of diverse and healthy ecosystems. Wildflowers, in turn, provide habitat and sustenance for various wildlife species.
- Fruit and Seed Production: Through their pollination efforts, bumblebees facilitate the production of fruits and seeds. This not only benefits humans through increased food availability but also supports a wide array of wildlife that depends on these fruits and seeds for sustenance.
- Biodiversity Support: Bumblebees are key species in the food chain. They help sustain diverse ecosystems by aiding in the reproduction of plants that serve as primary food sources for numerous animal species, ranging from songbirds to grizzly bears.
- Ecosystem Stability: Bumblebees contribute to the overall stability of ecosystems by promoting plant diversity and ensuring the availability of food for various organisms. The loss of bumblebees can disrupt these intricate ecological relationships.
- Human Dependence on Ecosystem Services: As humans, we rely on the services provided by healthy ecosystems, including pollination and the regulation of plant and animal populations. Bumblebees are integral to maintaining these ecosystem services.
- Conservation of Biodiversity: The decline of bumblebee species, including the rusty patched bumblebee, is indicative of broader biodiversity loss and ecosystem degradation. By protecting bumblebee populations, we contribute to the conservation of biodiversity as a whole.

In essence, the rusty patched bumblebee and its bumblebee counterparts are critical components of ecosystems that support both human and animal life. Their role in pollination and ecosystem functioning makes them indispensable, and their decline underscores the urgent need for conservation efforts to protect these vital species and the biodiversity they sustain.

7.7 Native Pollinator Ecology

The growing concern for the survival and well-being of bee populations, which has been evident since at least 2004, has led to the emergence of bee-friendly initiatives in cities and the establishment of pollinator gardens in many parts of the world.

Municipal policies have increasingly emphasized practices that support native pollinator ecology.[3]

These initiatives include:

- Cultivation of Native Flowering Meadows and Hedgerows: Promoting the growth of flowering meadows or hedges using native flowers and shrubs within urban areas to provide essential forage for native pollinators throughout the blooming season.
- Rewilding Techniques: Implementing rewilding techniques that aim to restore natural habitats within urban landscapes, creating more favorable conditions for pollinators.
- Green Manure Cover Crops: Promoting the use of green manure cover crops that enrich soil fertility while also offering nectar and pollen sources for pollinators.
- No Lawn Chemicals: Abandon the use of lawn chemicals that can harm pollinators and their habitats and develop an attitude of acceptance for dormant grass and brown patches. Sowing short varieties of clover will diversify a lawn and protect it against dry spells. Clover guarantees the longevity of the green color while offering pollen to pollinators and nourishing the soil. The short variety allows for easier mowing, but it is preferable to space out the mowings and keep the flowers blossoming.
- Limited Mowing on Private and Public Lands: Implementing limited mowing practices on both private and public lands allows wildflowers and native plants to flourish, benefiting pollinators. A bee-protection movement has started in North America, encouraging people to let the grass grow until June to protect dandelions, which serve as the first source of food for pollinators in the spring.
- Promotion of Pollen-Rich Plants: Efforts to encourage the use of pollen-rich plants in urban landscaping, both for the benefit of urban residents and pollinators. Some initiatives extend to include plants that support Monarchs, such as milkweed, as well as plants attractive to songbirds and hummingbirds.

Supporting invertebrate pollinators' health and survival is of paramount importance. Nalepa and Colla (2023) have proposed a comprehensive list of 83 solutions to achieve this goal. Some of these solutions include seeding wildflowers near plowed fields, leaving field edges unmown, and planting green manure crops or canola seeds between crops. This holistic approach is also advocated by Emma Marris (2011), who recognizes the value of nature wherever it exists, whether in residential backyards, hedgerows, road ditches, or along highways. These interstitial spaces serve as ecological connectors for various species within urban ecosystems, highlighting the importance of supporting not only bees but all forms of life. Even the smallest diversified units of nature offer ecosystem services to their surrounding environment and act as buffer zones against human impacts on declining

[3] Crops such as clover and buckwheat cover the ground in-between two harvest periods. Those two are both attractive for pollinators. At the time of preparing the fields for the next season, green manure crops are chopped and integrated in the soil to enrich it. This is where they derive their name from.

populations, such as songbirds and butterflies (Mader et al., 2011; Xerces Society, 2011; Little, 2016, Mortillaro, 2022).

In fact, the decline in songbird populations, as reported in the media, is attributed to similar stressors faced by bees. These stressors include habitat loss, pesticide use, and environmental changes such as altered land use and the consequences of climate change. For example, bobolinks nest in plowed fields in the spring and early summer in North America, but due to global warming, farmers have started to mow the fields earlier, resulting in the destruction of nests and young birds on the ground. Warnings are sent to farm owners, encouraging them to adopt practices favorable to the survival of this species. These practices include delaying hay harvesting until mid-July, changing the pattern of harvesting in the fields to a centrifugal movement from the center to the periphery to provide a safe passage for evasion, protecting a perimeter in the field until July 15 and around nests, adjusting the height of the implements used to avoid impacting nests, and more (UPA, website). The plight of disappearing songbirds further underscores the urgency of addressing these issues and implementing conservation measures.

Disappeared Havana Night Butterflies

Additionally, the decline in butterfly populations is a cause for concern due to their role as pollinators. An example from Havana, Cuba, illustrates how the population of large black butterflies has dwindled due to frequent fumigation efforts aimed at combatting mosquito-borne diseases like dengue, malaria, and the Zika virus (Weissenstein, 2016). Similar fumigation practices using kerosene/oil spraying and the introduction of predators (ex. cats) in tourist hubs like Cayo Santa Maria have adversely affected beneficial insects and songbirds, which tend to disappear despite the presence of an adjacent natural reserve (Aguilar, 2009: 3). While the Cuban government takes pride in having managed the Zika virus better than the rest of Latin America, have the other consequences on human health and the environment been assessed? These examples highlight the need for more sustainable and pollinator-friendly practices to ensure the survival of diverse pollinating agents and the preservation of native pollinator ecology.

In contrast to urban and touristic fumigated spaces, honeybee populations in the Cuban countryside enjoy good health. This serves as a testament to the beneficial effects of low-level chemical pesticide use in agriculture on the island. Cuban agriculture has effectively become organic by "default" due to the inability to purchase chemical products from the United States, owing to the trade embargo in place since 1962. Consequently, Cuban honey can obtain organic certification, a fact confirmed by the Cuban UN Food and Agriculture representative (Reuters, 2016; Renault, 2017). Cuban honey has garnered significant demand on European markets and has risen to become the "fourth most valuable export product" for the Cuban nation (Idem).

Between 2013 and 2015, I had the opportunity to visit various urban and peri-urban gardens in Cuba and to speak with several gardeners and academics. The most internationally recognized is the Vivero Alamar on the outskirts of Havana, which welcomes international visitors (Organoponico Vivero Alamar 2018 video). Cuba's unique political situation and island geography offer a valuable opportunity to study the far-reaching impacts of pesticides in agriculture on bee populations. The country's relative political and geographical isolation, particularly its limited trading partners and no immediate neighbors, has likely contributed to protecting its bees from contamination by external pathogens. Furthermore, the collapse of the Soviet Union in 1991 and the ensuing Special Period, which led to oil scarcity, had a transformative effect on Cuban agriculture (Wright, 2009). Prior to 1991, the Soviet Union provided subsidies for Cuban agriculture and was its primary trading partner, notably purchasing Cuban sugar cane production. In the wake of the agricultural restructuring that followed the post-1991 period, Cuba emerged as a global leader in urban agriculture (Castañeda Abad et al., 2017).

The development and adoption of various agroecological techniques in organopónicos (organic gardens) helped steer urban farming in Cuba, such as vermicomposting[4] to amend the soils, plant-based pesticides, and the use of raised garden beds to prevent soil erosion from rain. These innovations can be attributed to Cuban ingenuity and the imperative to ensure national food security (Wright, 2009; Padrón Padrón, 2015; Herrera Sorzano, 2018; Gamboa Costa et al., 2013). With over 900 urban gardens in and around Havana alone, Cuba's experience underscores the potential benefits of transitioning toward locally-grown produce through more sustainable and agroecological agricultural practices, benefiting not only the environment but also human health and food security (Gravel and Mestiri, 2015).

In North America, there is a consensus emerging in the urban planning practitioner's community around the importance of letting wildflowers and flowering meadows grow to increase "pollination reservoirs" near croplands and orchards. This approach is supported by various studies and resources, including those by Venturini et al. (2017), Everaars et al. (2018), and Ward and Adamson (2015). The US Department of Agriculture (USDA) Forest Service offers numerous publications on how to garden with pollinators, increase biodiversity in riparian buffer zones, and manage forests to support pollinators. Thinning forests to allow the undergrowth to be exposed to sunlight while leaving dead trees on the ground or standing can create ideal conditions for nectar and pollen-rich plants to bloom and create nesting habitats (Ward & Adamson, 2015; Colston Burrell, 2006; Leopold, 2005).

Additionally, flowering shrubs such as blueberry, elderberry, black raspberry, witch hazel, and hazelnuts are emphasized in these efforts, as they bloom at different times in the growing season, contributing to feeding pollinators during the spring, summer, and fall seasons (Ward & Adamson, 2015; Du Clos et al., 2022).

[4] The vermicomposting technique utilizes earthworms to decompose animal manure, providing an excellent soil amendment for use in flowerbeds. This technique allows gardeners to achieve seven times the productivity of a regular urban garden.

Planting flowering shrubs and leaving a flowering edge have been recommended to Indian orchards managers at the Himalayan foothill as part of pollinator recovery measures (Luthra et al., 2023).

A valuable tool for farmers has been developed by the University of Maine BeeMapper group: the InVEST Crop Pollination Model. This model predicts bee abundance in crops such as blueberries over a season, helping farmers make informed decisions to support and optimize pollination services in their fields. These practices and resources collectively demonstrate a commitment to creating and maintaining pollinator-friendly environments that benefit both agricultural productivity and the conservation of native pollinator species. The efforts of urban planners, researchers, and agricultural experts are crucial in promoting sustainable and pollinator-friendly land use practices.

To address the pressing need to create more hospitable environments for native pollinators, the focus of my research has centered on the quality of their habitats and the impacts of evolving farming practices on their populations, utilizing a landscape study approach. Qualitatively assessing the presence, density, and ecological value of pollen-rich plants in the landscape provides valuable insights into the continuum of landscapes, ranging from less to more favorable for native pollinators. This assessment can guide land-use management practices aimed at protecting and enhancing wild pollinators' habitats (Venturini et al., 2017).

Drawing parallels between past farming practices and dormant seeds left in the ground, we have an opportunity to derive inspiration from the past for contemporary approaches to farmland management. Insights into traditional know-how can help reorganize and rethink the relationship between farmsteads and their surrounding territories, aligning with sustainable food production principles. By integrating traditional practices and innovative solutions, we can create landscapes that are not only productive for humans but also hospitable and supportive of nonhumans, such as wild pollinators (Luthra et al., 2023).

The changing land-use patterns in rural history are primarily attributed to farm modernization, which is a major factor responsible for habitat loss for wild pollinators. Mechanized mowing and reaping, crop specialization, and the end of fallow periods have led to the loss of plant biodiversity, especially the disappearance of native species. This decline in biodiversity is a key factor in the reduction of pollinator populations. Now, we will examine two prominent trends observed in the North American countryside today: the expansion of monocrop landscapes and forest regrowth. Both have had detrimental effects on the availability of food resources, including pollen and nectar, for wild pollinators.

One of the key factors driving the depletion of food resources is the scarcity of native pollen-rich plants (PR plants) in these modernized landscapes. Additionally, there may be a significant increase in the distance between nesting spaces and the sources of pollen, which poses challenges for wild pollinators. For example, the largest of the native bees, the bumblebee, typically travels no farther than 1 kilometer (0.6 miles) to forage for food, and most other wild bee species stay within a radius of 500 meters (0.3 miles) from their nests (Ward & Adamson, 2015).

Wild bee ecology studies have demonstrated the impact of landscapes and climates on pollinator biodiversity. Nordic and temperate forest landscapes tend to support smaller populations of wild native bees, while regions with year-round abundant flowering plants and nut and fruit trees, such as Brazil (with 5,000 native bee species compared to 4,000 in the United States and 800 in Canada), provide a conducive environment for thriving wild bee populations (Agriculture and Agri-Food Canada, 2014; van Vierssen Trip et al., 2020; The Xerces Society, 2011: 26).

In regions with less biological diversity and fewer pollen-rich plants, such as northern Maine away from the coast, forests dominated by spruce and pine trees do not offer sufficient food resources for pollinators to thrive during the whole growing season. However, as we move southward through Central Maine, the landscape exhibits a more balanced ratio of evergreen and deciduous trees, including oak, chestnut (though disappearing), willows, poplars, and fruit trees. These diverse tree species provide a more favorable environment for wild native pollinators by offering a richer and more varied source of food, ultimately supporting their populations and well-being.

The concept of a "bee landscape," as proposed by the Xerces Society (2011: 94–95), envisions an ideal habitat for wild pollinators characterized by a multifaceted and interconnected ecological system. This holistic approach combines various elements, each thoughtfully designed to cater to the specific needs of these essential pollinators.

Within a "bee landscape," you would find a combination of features, including pollinator meadows ("bee pastures"), native butterfly gardens (combining native nectariferous plants with host plants for butterfly larvae, such as milkweeds) (Asclepias), hedgerows of flowering trees and shrubs, streamside and rangeland revegetation efforts, flowering cover crops, and pollinators' lawns (showcasing a mix of grass, native low-growing plants, and clover, for example). These components should be researched to identify local native plants and strategically integrated to provide a diverse array of resources and habitats for wild pollinators (Tallamy, 2018: 111).

One key consideration is the presence of an ample concentration of their preferred plants within a given area. Bees often select a single flowering plant and forage exclusively on it. To accommodate this behavior, creating "islands" or patches within the landscape with an abundance of the same wildflowers is essential for attracting and sustaining bee populations (Fig. 7.2).

The implementation of these practices considers the various requirements of bees throughout their life cycle, including foraging, overwintering, and nesting needs. This comprehensive approach aims to create a thriving "bee landscape" that not only supports pollinator health but also contributes to broader goals of biodiversity conservation, ecosystem health, and sustainable food production. It underscores the interconnectedness of pollinators with their environment and emphasizes the importance of responsible land management practices in their conservation.

Fig. 7.2 Wildflower Patch on a Lawn
Description: An abundance of dried golden rods can be seen in that wildflower patch left on purpose by the owner. This picture was taken at the beginning of October, that is at the end of the blossoming season in the province of Quebec, Canada. Maple tree foliage in the background is already entering in the fall season

7.8 New England Farming History: 1800–1920

While progressive innovations in farming techniques and tools had been emerging since the 1830s, a significant transformation occurred in traditional self-sustaining New England farms during the post-Civil War era, beginning in 1865 (see Tables 7.1 and 7.2). This transition was facilitated by the development of roads and railroad infrastructure, which enabled agricultural surpluses to reach distant markets (see Fig. 7.3). As early as the 1850s, larger and more specialized farm operations had started to emerge as a result of the technological revolution, leading to increased farm productivity. In Maine, farmstead owners embraced various market-oriented endeavors while preserving an open and multifunctional rural landscape (Maine Historical Society, 2000).

One notable example of this transformation was the flourishing dairy industry in rural Maine. Dairy production experienced a remarkable 500% increase between 1820 and 1850 (Horwitz, 1978: 38). The introduction of refrigerated train cars on the Bangor-Portland-Boston rail line in 1881 allowed Maine cheesemakers, many of whom also produced cream and butter, to expand their reach. This shift gradually

Table 7.1 Agricultural Innovations in Maine, 1820–1950

Change in farming practices	Description	Reason	Impacts on work/ landscape source	Source
Agricultural exports	1820s[a] Export apples (long conservation variety), flax seeds to Ireland	Family farming business opportunities	Specific crops from family orchards and croplands	Hardy (2008: 113–114)
Transformation of harvesting Implements	Mid-1830s	Farmers' preferences: time of execution, adjustment, posture	Abandoned the sickle, adopted the scythe; extended cropland area	Horwitz (1978: 60)
Mechanization of hay and grain harvesting	1850s Horse-drawn McCormick[b] hay mowers and reapers. More common use of machinery	Saved time, created long rows, easier to dry up the hay. Lessened farmers' dependence on paid workers	Crop fields extension. Decreased employment opportunities for farm boys	Falk (2012); David (1966) in Olmstead (1975: 327-28); Horwitz (1978:52)
Change in farm buildings and technology	1896 From haystacks to silos: silos and fed silage for dairy cows	Protected hay from the elements, preserved fodder quality, made it readily available all year round: boosted production	Disappearance of haystacks; improved capacity to store fodder; allowed dairy industry to develop	Maine Historical Society (2000: 6)
End of subsistence farming, replaced by commercial farming	1850s Decline of poly crops (cereal, corn and vegetables), specialization in the apple, wool and dairy production[c], sweet corn and potatoes in row crops, poultry[d]	Profit, business opportunity, hardship, improved access to markets with roads and train (arrived in 1849 in Winthrop, Me)	Expanded deforestation for pasture (animal feeding) and monocrop farming	Horwitz (1978: 38–39) and Maine Historical Society (2000)
Introduction of non-indigenous	1800–1850	Market-oriented monocrop agriculture. Introduced varieties: corn, apple, potatoes	Loss of ancient landraces[e] and biodiversity negatively impacted pollination reservoirs	Hardy (2008: 111–113)

(continued)

Table 7.1 (continued)

Change in farming practices	Description	Reason	Impacts on work/ landscape source	Source
Commercial farming development	After the Civil War (1865)	Railroads and industry developed	Market-oriented monocrop agriculture, loss of in situ biodiversity	Maine Historical Society (2000: 2)
Factory-transformed products for regional markets and exports	1871 Maine cheese production expanded. 1881 Refrigerated train cars	Refrigerated train cars allowed for expanding dairy products markets	Industrialization of agriculture: increased croplands used for fodder and monocrop production	Maine Historical Society (2000: 3–5)
Invention of metallic tools: from wood to iron	1882 Invention of metal blueberry rake helped make the industry profitable	Timesaving productivity	Blueberry patch expansion. Simplified workers' lives freed them from some manual chores. Increased working capacity per acre	Maine Historical Society (2000: 3)
Intensification and specialization of production	1879–1919 Emergence of crop chain value. Collective action called for defending crop specific farmers' interests by Government	Crop specialization seen as a way out of poverty for farmers but brought difficulties and dependence to monocrop system. Ex. 1928 hardship for potato growers	Obligation to grow or die. Farm abandonment impacted lives and landscapes. Contributed to rural exodus toward the city. Forest regrowth	Maine State Grange (1928: 70)
Advent of automobile and highways	1914–1935 Maine entered the automotive age	Transportation needs for commodities and people	Hay production decreased around cities with urban sprawl, loss of cropland	Falk (2012)
Out migration of farm workers to the West (Ohio)	1930s Cold summers, lost crops		Farm abandonment, forest regrowth, impacts on bees: habitat and pollen loss	

(continued)

Table 7.1 (continued)

Change in farming practices	Description	Reason	Impacts on work/ landscape source	Source
End of summer fallow	1970s	Increased acreage of cropland, productivity, chemical fallow as substitute	Monocropping; use of herbicides allowed for ending or shortening summer fallow periods, impacting negatively in-field biomass and farm ecosystems	Wicks and Simka (1973) and Agriculture Canada (2014)

[a]Note that Maine became a state in 1820.
[b]Mechanized mowing with McCormick horse-drawn machine was invented in 1833–1834 but not adopted by farmers till 1850s. Reasons of this delay include high costs of buying, operating, reduced lifetime of machinery at first. Western states adopted it earlier in the U.S. and farmers regularly made a group purchase (Olmstead, 1975).
[c]Dairy production increased in 1850 by 500% compared to 1820 (Horwitz, 1978: 38).
[d]"Dried and ground at the mill, the homesteaders' corn provided a "meal" or flour that was the centerpiece of the family diet. The surplus was easily sold, and the waste was fodder for cows and oxen. In 1800, it was still grown in hills, instead of rows, along with beans and pumpkins, though row planting would become the norm in the next generation" (Hardy, 2008: 112).
[e]Indian maize declined, grown in Wells (Southern Maine) till 1818.

led to the abandonment of butter and cheesemaking activities within domestic spheres, freeing up the female labor force. Many women joined the workforce at cheese factories or textile mills in southern Maine thereafter (McMurry, 1995).

As a result, the self-sustaining farmstead economy became increasingly precarious, compelling farming families to continuously adjust the scale of their operations, diversify their crops, and allocate varying amounts of labor to home product transformation. During this period, town retailers began offering ready-to-use consumer goods, providing the much-needed time-saving solutions for busy homesteaders.

7.9 Premodern New England Farmstead: A Hub of Diverse Functions

The Historical Atlas of Maine, authored by Hornsby and Judd (2015: 52), includes a map depicting the major crops by region in 1850, with an emphasis on the crops with the highest volume of production. These crops included corn,[5] oats, potatoes, fruits and vegetables (particularly apples), wheat, and buckwheat. Rye, peas and

[5]Notably, there were 100 corn canneries in existence after the Civil War, but corn production is now nearly nonexistent (Idem).

Table 7.2 Early Nineteenth Century Farmstead Characteristics (1800–1850)

Agricultural function or feature	Description	Reason	Impacts on work/ landscape	Source
Homestead layout and vegetation	Biodiverse fields and farms with fruit trees, lilacs, sugar maples in front of the house + kitchen garden	Family's enjoyment and subsistence	1/3 in woodlot, the remainder in pasture and tillage. Crops: corn, oats, potatoes, flax[a], barley, wheat, buckwheat, potatoes, and pasture, cow alley	Hardy (2008: 110–111)
Multifunctional and biodiverse homestead	Kitchen garden, subsistence crops, hay, livestock, crop and milk, and wood	Family subsistence	Surplus sold to markets (cream, butter)	Hardy (2008: 109–113) and Maine Historical Society (2000: 3)
Soil amendment	Natural fertilizers such as ash, fish (alewives), cow manure, rockweed, or seaweed (brought from the shore)	Manure added before planting to enrich the soil	Natural manure used locally available material, no energy lost on the farm. Labor intensive	Hardy (2008: 111)
Cattle feed	Potatoes and corn ears	Both potatoes and corn were "multipurpose crops" for human and animal feeding	Self-sufficient farmstead, waste recycling	Hardy (2008: 113)

[a]Flax seeds were a major export to Ireland for linen production. Also used locally for linseed oil and linen making. Hardy (2008: 112–3)

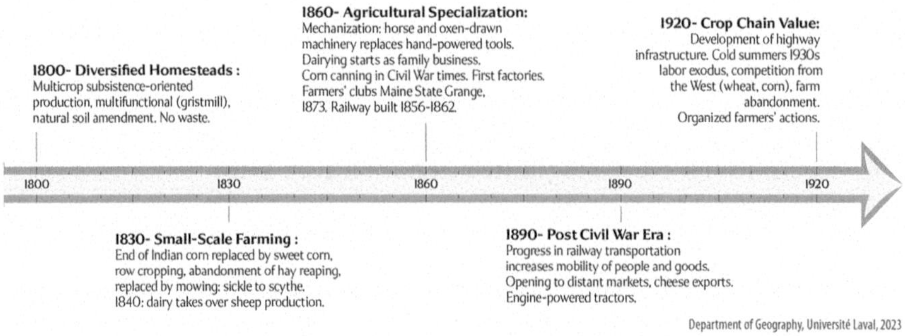

Fig. 7.3 Timeline of Changing Agricultural Practices in Maine, 1800–1920

beans, flax, hops, and hay completed the list. In the Kennebec Valley region, corn and oats were the most abundant, while in the Southern region, corn and potatoes dominated the agricultural landscape.

Fig. 7.4 An 1863 Winthrop Farmhouse with an Apple Tree in its Front Yard. (Credit: Gravel, 2019)

Historic New England farmsteads (1800–1850) were hubs of diverse functions, each serving a specific purpose within the agricultural ecosystem (see Table 7.2). The layout of these farms was typically divided into distinct tiers, with approximately one-third dedicated to woodlots, one-third to pastureland, and one-third to cropland. Additionally, the farmstead featured essential elements such as a garden, an inviting front yard adorned with a tree alley (often featuring sugar maples), and gardens filled with vibrant flowers (Sanford, 2015: 64; Horwitz, 1978; Hardy, 2008).

The sugar maples in the front yard served a practical role, providing sap for the family to transform into maple syrup during the spring. In close proximity to the family kitchen, a small vegetable garden thrived, supplying the cook with fresh produce. Moreover, a small orchard with apple trees complemented the garden (Fig. 7.4). Together, these elements allowed the farmstead to meet most of the family's needs, encompassing wood for heating, a variety of foodstuffs (including dairy products, meat, vegetables, and fruits), and storage spaces for food preservation methods such as drying and canning. This self-sufficiency enabled the family to enjoy a diverse array of food throughout the winter months.

The rich tapestry of functions performed by a nineteenth-century farmstead provided its residents with a degree of independence from the outside world. The list of farm buildings reflects the wide range of activities undertaken by its members, including blacksmithing, greenhouse cultivation, stable and carriage maintenance, milk processing, cheese production, kiln operation, wood storage, beekeeping, hop cultivation, corn storage, long-term crop storage, manure management, poultry husbandry, grain storage, silage storage, sheep care, maple syrup production, meat preservation, root cellar storage, springhouse utilization, and ice storage (Sanford, 2015: 73). It's worth noting the presence of beehives within this diverse array of farm functions, highlighting the importance of bees and honey production in sustaining the farmstead's ecosystem.

Sheep and apples played complementary roles in this agricultural framework, although apple orchards were not initially widespread during the early years of New England's colonization. Most farmsteads had just one apple tree, primarily to provide for the family's consumption, with little surplus for the market. A tradition of slicing and drying apples allowed families to preserve and regionally market this fruit (Nelson, 1945: 167). Florence Agnes Nelson, reminiscing about her childhood on a Maine farm, recalls the communal activity of slicing apples by hand (1945).

As commercial orchards began to develop, sheep were introduced to these orchards, serving multiple purposes. They helped fertilize the soil with their grazing activities, consumed windfall apples, and acted as natural weed control. This integrated approach is now recognized as an "agrosylvopastoral system" by agroecologists. At times, sheep were replaced by hogs (Bowler et al., 2012). The synergy between various elements within these systems resembles the humanized landscapes of the Spanish and Portuguese Montado (Portugal) or Dehesa (Spain) agroforestry systems, where black Iberian pigs are raised beneath cork oak forests. The inclusion of acorns from native oak trees in their diets imparts a distinctive flavor to the resulting pork meat, reflecting the terroir. In the Montado/Dehesa system, animals play a crucial role in maintaining both the ecosystem and the landscape.

Comparing the traditional New England farmstead with the Mediterranean farmstead underscores the importance of local knowledge and wisdom in redefining the relationship between humans and nature. These examples serve as a source of inspiration for reimagining sustainable agri-food systems that harmoniously coexist with the environment, emphasizing the interdependence of various elements within the ecosystem.

The apple industry began to flourish in Central Maine during the early twentieth century, and the town of Monmouth was particularly blessed with favorable growing conditions. By 1950, the Chick family had established a prominent apple sales and processing company, boasting ownership of 23 individual orchards with a total of 26,000 apple trees spread across three different towns—Monmouth, Winthrop, and Greene. Their apple products were exported to Europe and various parts of the world, while locally and in Boston, they were marketed under the well-known "Chick-a-Dee" brand. The McIntosh variety of apples was highly favored for its ability to withstand travel and preservation, despite the considerable need for spraying to combat fungal diseases (see Fig. 7.5) (Bowler et al., 2012: 52, 65). Notably, during this period, the pollination stocks and services in the area were substantial, as apple trees heavily relied on bee pollination for fruit production.

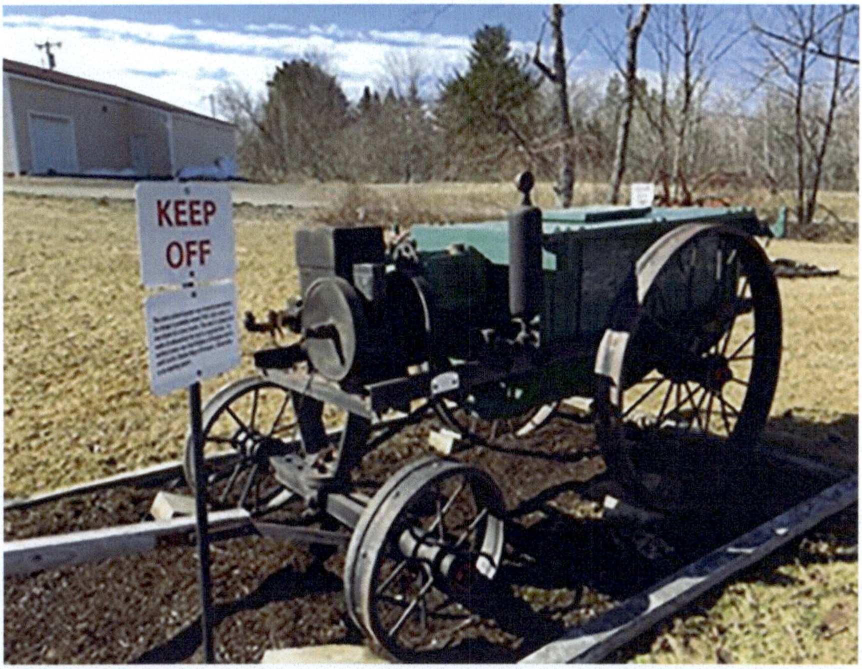

Fig. 7.5 Old Orchard Sprayer, 1940, Orrington, Me. (Credit: Gravel, 2019)
Description: This photo was taken at the Wiswell Outdoor Farm Museum in Orrington, Maine. The sign in the image provides historical context: "This early orchard sprayer was drawn by two horses. Its vintage is probably 1940. In recent years, it was drawn by a farm tractor. The tank is original and made of redwood, but the top is a reproduction. Its earliest owner was Fred Glidden of Orrington, and he sold it to Dr. Charles King of Orrington. Restoration is an ongoing project"

7.10 Rural Modernization and Biodiversity Loss on the Farm

The first half of the twentieth century ushered in a new set of challenges, requiring adaptation to economic and climatic fluctuations. For instance, during the 1930s, the region experienced unusually cold summers, leading to a significant loss of farm labor as farmworkers sought alternative livelihoods in industries or urban areas. Some even ventured west to states like Ohio, where more favorable growing conditions existed. Their success in these new ventures soon became apparent, flooding Atlantic seaboard markets with wheat and other grains and produce from the wheat basket states. This influx led to a decline in family agriculture in Maine, with only niche markets for dairy products, fodder, select orchards, and blueberry production, primarily in the Downeast region, remaining.

Some reports attribute the loss of farmland biodiversity to developments dating back to the 1940s, driven by efforts to straighten rivers, the reduction of fallow land, the widespread use of pesticides, and the expansion of monocropping. These changes were aimed at increasing cropland productivity for post-World War II market-oriented production (Agriculture and Agri-Food Canada, 2014). However, this research suggests that it is beneficial to revisit the earliest stages of agricultural modernization. This includes the transition from manual or horse-powered tools, such as the scythe used for manual grass cutting, which was later replaced by horse-drawn mowers and eventually tractor-mounted mowers. The advent of engine-propelled mowers and tractors in 1939 marked a significant shift in agriculture, granting farmers greater control over their land in terms of both work intensity and field size (Falk, 2012). These tractors and mowers, which replaced animal-drawn technology, empowered humans to exert a more profound influence on their environment. They enabled the comprehensive clearing of vegetation, deeper soil tilling, unprecedented soil compaction, and extensive tree cutting without distinguishing between timber resources and other plants or trees (Olmstead, 1975; Daum, 2023).

This transition resulted in a more uniform landscape shaped by a productivist perspective that prioritized resource-oriented development for the sake of growth and profit. The traditional tiered land pattern, with one-third allocated to woodlots, one-third to farmland, and one-third to gardens and orchards for self-sustenance, gradually vanished (Hardy, 2008). In the 1880s, the emergence of regional markets and increased competition from Western farmers prompted Maine farmers to seek greater efficiency and specialize in the most profitable activities their farms could support. This shift toward specialization led to a reduction in crop diversity in favor of more specialized endeavors, such as dairy farming, with a primary focus on producing animal feed, particularly hay. While specialization held the promise of increased profits, it also made farmers more susceptible to fluctuations in market prices, not to mention the economic and agronomic risks associated with monocropping.

Challenging years marked by cold summers, crop failures, and blights further incentivized labor outmigration and the closure of many farms. During this period, cities began to attract young rural individuals with promises of employment and educational opportunities, including the emergence of young female laborers in textile plants in places like Wells (McMurry, 1995; Hardy, 2008). Some farmers even reminisced that 1820 was the last truly profitable year for Winthrop farmers in Kennebec County (Horwitz, 1978: 37).

During this transformative period, as family farmsteads evolved into agricultural businesses, the capitalist economy was experiencing significant growth. Many of the family's essential needs, such as food, tools, clothing, and work equipment, which had previously been crafted using local resources, began to be sourced from the market. An examination of homemade supplies and farming tools during this era reveals a shift in craftsmanship, marked by the transition from traditional materials like wood to modern materials such as metal, glass, corning ware, and plastic. For example, wooden measuring cups gave way to metallic cups, and eventually, glass and plastic versions became commonplace.

This transition had a profound impact, as the time and effort previously dedicated to crafting everyday objects could now be redirected toward more productive activities. This shift reflected the broader trend of industrialization and the increasing availability of manufactured goods, which transformed both rural and urban life during this period.

Native seed varieties, such as Indian corn, followed a similar trajectory as handcrafted objects, gradually fading into obscurity. Indian corn, for instance, was cultivated until as late as 1818 in Southern Maine and on islands in Massachusetts. Some remnants of Indian cornfields were still identifiable until around 1930, as a few planting grounds managed to survive the European settlements that replaced them (Sanford, 2015: 46).

As agriculture evolved, exotic plants were introduced, some of which became invasive species, further reducing the space available for native species to thrive. One notable example is the dandelion, which was introduced in Boston in the 1770s and has since become a widespread invasive plant (Sanford, 2015: 67). This shift in the ratio of native to nonnative plants has had detrimental effects on wild pollinator populations, as these pollinators exhibit a strong preference for native plants that have coevolved with them over a longer period of time (Tallamy, 2012: 279–80).

The decline in native plant species has had negative consequences for biodiversity and ecosystems, emphasizing the importance of preserving and restoring native vegetation to support pollinator populations and overall ecological health.

The establishment of railway networks played a crucial role in connecting rural towns and territories, facilitating the dissemination of technological innovations from urban centers to remote regions. These innovations in agriculture, often driven by scientific advancements, were embraced under the banner of progress and time-saving techniques.

During World War II, a new wave of innovations and terminology emerged to combat various perceived "enemies," including insects and weeds. The battle against pests and unwanted vegetation received significant attention, leading to the publication of gardening booklets in the post-World War II period. These publications catered to both farm owners and Victory Gardeners, drawing on the scientific knowledge of the time. However, the guidance provided on the use of chemicals sometimes lacked the caution and prudence required.

The wartime era witnessed a surge in agricultural innovation and practices aimed at boosting productivity and supporting the war effort. These changes had a lasting impact on farming and rural communities.[6] From the early 1900s to the 1950s, extensive debates and discussions revolved around the modernization of rural areas and the utilization of natural resources, including harnessing the energy of rivers. Administrators of the Maine State Grange were acutely aware of the potential impacts of modernity on the long-term sustainability of rural regions. In fact, they issued prescient warnings well ahead of their time, as documented in the 1930 general annual meeting account (Maine State Grange, 1930).

Farmers' almanac editors also played a role in advocating for farm sustainability by offering guidance on various practices. They provided wisdom on how to manage grass on farmland without cutting it too short and emphasized the importance of

[6] The Prudence Principle is one of the 12 principles of sustainable development, advocating for the cautious use of powerful technologies until their full impact on human health and the environment is thoroughly understood.

treating farm animals with kindness and respect, recognizing that animals could not respond with words (Maine State Grange, 1896).

Despite these efforts to promote sustainable farming practices and humane treatment of farm labor and animals, the prevailing narrative of modernization, driven by scientific knowledge and progress, often dominated discussions on rural economic development. This emphasis on modernization was motivated by the pressing need to combat rural poverty. In doing so, rural schools and farmers' organizations received support from the Maine State Grange to promote rational approaches to farm management and governance, with the goal of enabling every member of the farm community, including men, women, and children, to realize their full potential both at home and as landowners or farmworkers (Maine State Grange, 1928).

During the 1930s, a period marked by environmental and economic challenges, farmers across the nation, united under the National State Grange, called for federal financial support to help them navigate these hardships (Horwitz, 1978: 39). This collective effort reflected the determination of farmers to secure their livelihoods and agricultural sustainability during trying times.

7.11 Tilting Point: Vanishing Farms and Forest Regrowth

The turning point in the agricultural landscape came with the decline of farms and the regrowth of forests. As competition from newly opened croplands in the Western United States intensified, New England farmers found themselves facing economic challenges. It became more cost-effective to import potatoes and grains from states like Ohio, Idaho, and Indiana than to continue producing them locally (Horwitz, 1978). Hardy's research reveals a significant decline in wheat production, with the wheat crop in 1850 only reaching 14% of its 1838 levels (Idem, 1978: 38).

Many Maine farmers shifted their focus from subsistence farming to commercial agriculture, which exposed them to a new threat. Between 1831 and 1836, inflation caused the prices of essential goods such as flour, butter, beans, hay, and firewood to more than double (Idem, 1978). The reliance on sheep as a source of income also faced a harsh end, as diseases wiped out entire flocks in Winthrop between 1828 and 1832. At one point in 1839, in certain towns in Vermont, New Hampshire, and Maine, there were more sheep than people! In other parts of New England, strong sheep production continued until the late 1840s, after which dairy cows began to dominate (Sanford, 2015: 42). By the 1860s, Winthrop had developed a reputation for having the finest dairy herd in Maine (Ibid. 2015: 42).

In 1850, the US. census reported that 32% of the land in Maine was utilized as farmland, in contrast to New Hampshire with 63% and Vermont with 84% (Idem 2015: 18). Dairy farming became a prominent source of income for residents in the Down east and Central regions of Maine, while blueberry plantations were established on the northern marginal coastal lands. This shift in agricultural practices contributed to the transformation of the dominant landscape in Central and Southern Maine over the past two centuries, characterized by extensive forest regrowth and a continuous decline in biological diversity.

Notably, the forest cover in New England in 1850 was estimated to be around 20%, a situation that has significantly reversed today (Idem, 2015: 17; Hornsby & Judd, 2015: 52). By 2015, Maine had become the most forested state, with approximately 90% forest cover, followed by New Hampshire at 84% and Vermont at around 78% (Sanford, 2015: 78). Over time, the number of farms in Maine reached its peak in 1880 and then began to decline, mirroring a trend that occurred across the rest of New England and Canada during the late nineteenth and early twentieth centuries. This decline in farming activity and the abandonment of farms were particularly notable in western regions of Maine, areas near the Appalachian Mountains, and on hilly and rocky soils. However, the construction of railways enabled the establishment of new farms in eastern Aroostook County, close to the border with New Brunswick, Canada, where the land was seen as promising for potato cultivation (Hornsby & Judd, 2015).

Inhabited forests have become a prominent feature of the Maine landscape due to the significant conversion of "improved lands" to forest regrowth since 1910 (Idem, 2015: 53). This way of life involves living in forest communities where residential houses are nestled within well-preserved natural environments. Simply driving on the backroads of Down east and Central Maine reveals how the forest's edge is gradually encroaching upon residences, with pastures being replaced by extensive forested areas. In some cases, these forests have grown to the extent that they completely obscure houses from view along the road (Fig. 7.6).

Fig. 7.6 House Nestled Within a Lush Forest in Winthrop, Maine, with a Cleared Roadside for Utility Purposes. (Credit: Gravel, 2018)

While this land-occupation pattern can provide valuable buffers against riparian erosion, reforested lands of this nature may also develop into large, homogeneous forest patches. Such patches often lack fragmentation and plant biodiversity, which can diminish the availability of pollen for bees. The Xerces Society has suggested that fire-managed forests could potentially help reopen habitats for pollinators. However, the application of this technique must be approached with care, as there is a risk of harming fauna and insects that should be protected, and it requires careful control to prevent wildfires from spreading uncontrollably (The Xerces Society, 2011: 94).

Stone walls and rock piles bear witness to the agrarian past, still evident alongside rural properties. These stones were painstakingly cleared from the soil to make way for agriculture, either forming mounds eventually covered by vegetation or used to construct walls for land parcel demarcation (Sanford, 2015; Hardy, 2008). Winthrop and Tatnic Hill in Southern Maine provide numerous examples resulting from farm abandonment, showcased in Fig. 7.7a, b. The Tatnic Hill Nature Conservancy preserve pays tribute to the "good stewardship" of past landowners with a commemorative plaque at its entrance. Exploring its trails offers scenic vistas and glimpses of house foundations and stone walls. Today, ospreys, partridges, and turtles thrive in this natural refuge that once was a working farmstead. Despite the arrival of younger generations, Tatnic Hill retains its historical landmarks, including a family Civil War cemetery at a crossroad, an old school surrounded by century-old trees, abandoned agrarian tools along crop field borders, old orchards, and fallow fields.

Climate vagaries played a significant role in farm abandonment in the past. For instance, Winthrop's orchards were devastated by bitter winters in 1831–1832, while the cold temperatures and drought of 1836–1837 worsened an already severe

Fig. 7.7 (**a** and **b**) Tatnic Hill Field with Orchard in the Background and Remnants of the Old Tatnic Road Stone Walls, Wells, Me. (Credit: Gravel, 2019)
Description: An abandoned two-wheel tractor is displayed in 7a. Note that this type of farm implement was animal-traction drawn. On the rightside of the tractor, vertically, can be seen as a tool to create rows. In Fig. 7.7a, b the old stone wall separates a hay field from an abandoned field, now the site of forest regrowth

situation of grain and fuel shortages. Similar conditions recurred in 1841 and 1849 (Horwitz, 1978: 38). These challenging weather patterns prompted adjustments at the farmstead level and contributed to the "Ohio fever," which led many farm owners and boys to leave New England. Consequently, maintaining plowed land became increasingly difficult.

As abandoned plowed fields lay fallow, they gradually transitioned into secondary forest growth. In the initial years, these fields were overtaken by weeds, grasses, and shrubs like raspberries. By the third year of abandonment, they transformed into moss-covered meadows featuring wildflowers and young trees, including willows, white pines, and poplars (common in New England). This shift from cultivated fields to secondary growth forests has left its mark on the contemporary landscape.

As seen in Chap. 6, in Northern Maine, the timber industry played a significant role in deforesting the Bangor region by 1853, a fact documented by Thoreau during his journeys to the Maine woods (Thoreau, republished 1950). These cleared lands were briefly used for cultivation but were eventually abandoned due to labor shortages and unfavorable market conditions.

This extensive forested landscape now poses a threat to what Venturini et al. (2017) referred to as "pollination reservoirs." Without the openings created by backroads and power line corridors, the proliferation of dense, often coniferous, forests negatively impacts the survival of wild pollinators. Forest fragmentation, on the other hand, provides more favorable habitats for wild pollinators, including blueberry patches managed with controlled chemical sprays or prescribed fire,[7] as well as orchards (Venturini et al. 2017; Du Clos et al., 2022). This underscores the importance of sustainable forest management practices in supporting wild pollinator populations in the region.

7.12 Beescaping Practices

Pollinators tend to do well in disturbed land areas, such as along forest roads, ditches with abundant wildflowers, or in deforested areas during the first 3 years of regrowth, when raspberry bushes and flowering meadows appear. Urgent action is needed to offer pollinators feeding and shelter options within forests and fields, as well as chemical-free living spaces to address "pollination-insecurity" (Venturini et al., 2017). Many wild and solitary pollinators forage in soil, with approximately 70% of them preferring sand or gravel material. The remaining pollinators live in natural cavities, such as trees, or in manmade structures like fences, where they establish

[7] Indigenous groups living on coastal lands in Maine used the fire to burn the land and grow blueberries. The following year, the surviving blueberry roots gave higher yields. The controlled fire technique tends to be replaced today using chemical spraying, but some traditional growers still use it for the social gathering, the tradition and the quality of the produce (Fater, 2020).

residences for laying eggs and feeding their larvae (Hunter & Lightner, 2016; Normandin et al., 2015). The intricate nature of plant–pollinator interactions at a fine spatial scale necessitates a landscape approach focused on habitat type and ecological value. As part of the "reconciliation ecology" trend (Rosenzweig, 2003) calling for the "redesign of human habitats for the accommodation of other species" (Tallamy, 2018: 37), McFrederick and LeBuhn (2006) proposed the concept of a "reconciliation landscape" to provide native pollinators with easy access to their preferred habitats.

Studies have demonstrated that forest regrowth following fires or cuts, such as maintenance tree cuts in power line corridors, can be beneficial. Raspberry and blueberry bushes are among the first vegetation to emerge in such areas (Du Clos et al., 2022). The diversity of blossoms in power line corridors further enhances their appeal as habitats for wild bees (Idem). It's important to note that this is not an endorsement of clear-cutting forests; rather, it highlights the potential for landscape fragmentation to mimic old farmstead patterns or humanized landscapes, which could reintroduce more agricultural functions to rural counties.

Many authors, including Venturini et al. (2017: 108), suggest that the ideal farmland should include 1/3 natural habitat to preserve pollination reservoirs.

> When 25–30% of the landscape within 750–1000 m of the crop field is natural habitat, 100% of crop pollination needs were met in watermelon fields according to Kremen, Williams, and Thorp (2002), the profitability of canola was significantly increased (Morandin and Winston, 2006), and bee populations were maintained over time (Banaszak, 1992; Venturini et al., 2017: 108).

While the exact percentage of pollen-rich plants needed in the overall "natural habitat" remains uncertain, it's crucial to prioritize the removal of exotic invasive plants to allow indigenous plants to thrive. Indigenous plants are more likely to be recognized by wild pollinators if they offer pollen and nectar. Many native bushes produce flowers and fruits that can attract wild pollinators and birds, making them ideal choices for planting. With readily available identification tools, individuals can become rewilding agents in their local environments. The best time to collect and sow seeds outdoors is in the fall when they are dried on the branches. Care should be taken to create pollinator corridors.

Even a small gain, such as adding 1–5% of pollen-rich plants alongside cropped areas, can boost lowbush blueberry field production by approximately 10% 3 years after establishment (Venturini, interview, 2018). Pollinators benefit most from dense patches of recognizable floral resources.

While some dandelions and raspberry bushes may be expected to grow during the warm season, bees require a complete blossoming season to sustain themselves on pollen (The Xerces Society, 2011). This sustenance is made possible when various plant species with different flowering times throughout the summer contribute to the pollen stock.

7.13 Reviving Maine's Farming Tradition: Feeding Humans and Nonhumans

As Maine's current farmers approach retirement, more than three-fourths of the state's farmland is poised for potential transition or change in occupation (Edwards, 2017). Nongovernmental organizations like Kennebec West are actively supporting young farmers who aspire to start their careers in the area. They are also encouraging existing farmers to remain in the profession by advocating for towns to implement tax relief programs, as is the case in Winslow.

Kennebec West, in collaboration with the Manchester Conservation Commission, has adopted a regional approach that promotes cooperation among neighboring towns, including Winthrop, Monmouth, and Manchester. Their shared goal is to find solutions to the ongoing loss of farmland to urban development and transportation infrastructure. By reducing the tax burden on farmers, they argue that farms do not place as much demand on municipal services as residential developments do, humorously summarized by the phrase "cows don't go to school." This tax relief program can provide sufficient incentives for farmland protection against residential development, with the condition that the protected land remains as farmland for a specified period, often up to 20 years (Idem).

The Maine Organic Farmers and Gardeners Association (MOFGA)[8] aligns with these principles, with a particular focus on providing guidance and support to small, diverse farming operations. According to a report on the state of agriculture in Maine published in 2019 (Idem), MOFGA's contribution is noteworthy. While Maine has seen a decline in the number of farms overall, the silver lining is the growth of small organic farms. These farms, characterized by their diversity and modest sizes, have the potential to play a pivotal role in establishing the patchwork landscapes that are essential for creating vibrant and diverse pollinator habitats.

At the scale of a bee's foraging range, it is crucial for landscapes to offer a mix of features, including woodlands, polycropping, flowering meadows, wildflower hedges, and fruit trees. Intercropping, which involves planting alternating strips of vegetables and fruit trees among other crops, is an effective way to achieve this diversity, as opposed to monocropping practices. Additionally, integrating livestock such as cattle and sheep into these landscapes can further enhance their ecological balance. Cattle can help manage windfall fruit, while sheep serve as effective weed eaters, ensuring orchards remain free of invasive weeds and tree sprouts (Sanford, 2015: 51).

Despite the challenging trends in agriculture today, there is a notable "renaissance" or back-to-the-land movement in Maine (Curtis, 2019) that is attracting young individuals to farming. This movement is characterized by an emphasis on

[8] MOFGA, founded in 1971, plays a central role in promoting and supporting this back-to-the-land movement. As the oldest and largest state organic organization in the United States, MOFGA provides visibility to this resurgence in farming and has the potential to help reverse the trend of farm abandonment that was prevalent in Maine until 2017.

organic farming practices, a unique lifestyle, and a commitment to environmental values. It holds promise for Maine's agricultural sector and offers an opportunity for farmers to advocate for the conservation of farmland and the restoration of biodiversity.

Data from the 2017 US Census of Agriculture, covering the period from 2012 to 2017, reveals several significant trends in Maine's agriculture (Curtis, 2019: 3):

- A reduction of approximately 10% in farmland, with acreage decreasing from 1.45 million to 1.3 million acres.
- A decline of 573 farms over 5 years, from 8,173 farms to 7600.
- A 15.8% decrease in the average net income per farm, dropping from $20,141 in 2012 to $16,958 in 2017.
- An increase in the number of farms exceeding 2000 acres, indicating larger landowners are acquiring neighboring farms while the latter are facing dismantling.

These statistics underscore the importance of initiatives like MOFGA and the back-to-the-land movement in revitalizing Maine's farming landscape and preserving its agricultural heritage. On the other hand, positive trends reflect a promising future for Maine's agriculture, with growing interest in small-scale, direct-to-consumer, and organic farming, as well as a more diverse and youthful farming community.

Despite the efforts and dedication of farmers in Maine, challenges remain as they strive to secure their farms for future generations. While young farmers are encouraged to continue their family farming traditions or embark on new agricultural ventures, there is a concerning trend in the aging of Maine's farming population. The average age of Maine farmers has increased from 55.1 years old in 2012 to 56.5 years in 2017, according to the Maine Farmland Trust (Idem).

This aging demographic, coupled with the loss of farmland to various factors, paints a somewhat gloomy picture for the future. Many older farmers find themselves in a difficult position, as they may lack retirement options if they are unable to sell their farms. Moreover, they may be hesitant to sell their farms to residential developers, which could lead to the transformation of their agricultural land into forests or housing developments.

The challenge of preserving Maine's agricultural heritage and ensuring a sustainable future for farming in the state requires innovative solutions and support for both established and aspiring farmers.

The small farming and organic movement in Maine promote the idea of building community networks and support systems to achieve a level of autonomy and resilience. This approach emphasizes the importance of local self-sufficiency and the capacity of communities to navigate challenges and thrive independently.

Collaboration among neighboring communities at the regional level is proving to be a valuable approach for creating opportunities and addressing rural multifunctionality. This intercommunal collaboration is especially relevant in the context of organic farming. MOFGA's vision and mission align with the goal of fostering diverse and interconnected human communities.

As stated on MOFGA's website, their vision is "a future of healthy ecosystems, communities, people, and economies sustained by the practices of organic agriculture." The organization's mission emphasizes its role in educating and advocating for organic agriculture, highlighting its interconnectedness with a healthy environment, local food production, and thriving communities.

This spirit of collaboration and shared values can bring together farmers of all ages, enabling them to work toward building sustainable communities. It also provides an opportunity for farmers to revisit and refine their own criteria while making the transition from conventional to agroecological agriculture. By working collectively and embracing organic practices, these farmers contribute to the well-being of their ecosystems, communities, and the overall health of the region.

What implications does this have for rural landscapes? We can anticipate that a revitalized and multifunctional local rural economy will not only preserve open landscapes but also enhance floral diversity to support thriving pollinator populations. The restoration of abandoned farmland offers a ray of hope for creating more biologically diverse regions that can sustain both human communities and bee populations.

I concur with Sanford's observation that the current emergence of the local food movement within sustainable agriculture may bring about a renewed focus on traditional crops (Sanford, 2015: 49–50). Of the remaining orchards, old apple and plum trees, with proper pruning, can be coaxed back to bloom, thereby reviving ancestral varieties and providing vital pollen for pollinators from their flowers. For small farmstead owners who prioritize ecological diversity, this is one of the ways to rejuvenate historical rural assets and establish havens for native bees (Gravel, 2017, 2019). Recognizing the value of old outbuildings or natural attributes, such as fruit trees, can make a significant contribution to diversifying a farm's production at minimal costs (Gravel, 2004).

The advantages of farmland restoration are manifold and include significant benefits such as climate change mitigation, soil enrichment, and protection. Additionally, the establishment of tree and shrub plantations contributes to increased biodiversity and provides essential ecosystem services, particularly pollination (Yang et al., 2020). In many regions where forest regrowth has been a predominant trend in recent decades, there has been a growing interest in non-timber forest production and harvesting. These activities encompass a wide range of products, including berries, nuts, medicinal and dietary plants, and edible items like garlic, mushrooms, and various wild plants to harvest such as fiddleheads[9] that can be sustainably harvested.

In essence, the resurgence of rural economies with a focus on multifunctionality holds the promise of maintaining diverse and thriving landscapes, benefiting both

---

[9] Since some wild plants are endangered, it is best advised to check with one's own jurisdiction to see their status. It is always mindful to harvest in a way to allow the plant to survive and regenerate and to avoid depleting important patches, also thinking about leaving enough on-site for the following pickers.

local communities and the natural environment, including vital pollinators such as bees.

Agroforestry techniques present diverse models for tending to or managing forested lands, enabling increased yields while fostering interactions within forest ecosystems. To support the development of forest communities and revitalize abandoned farmland, municipalities and civic organizations must play a crucial role. Integrating edible and melliferous plants into these areas can promote biodiversity and provide numerous benefits (Chamberlain et al., 2020).

The establishment of farmers' market networks is vital for providing selling opportunities for forest and farm products. This is well illustrated by Silber (1988) in her account of transitioning to rural life as an urbanite in her book A Small Farm in Maine. However, it is equally important to have active involvement and support from stakeholders at various levels to finance and incentivize the initiation of forest farming ventures.[10] By facilitating cooperation and collaboration among these stakeholders, we can nurture a thriving forest-based agricultural sector that benefits both communities and ecosystems.

7.14 Imagining the Agroecological Transition: Envisioning Sustainable Agriculture and Community Resilience

The agroecological approach celebrates the interconnectedness of all natural elements on the farm, including humans, and seeks to harness the natural self-regulating and healing potential of ecosystems. Similar to how Indigenous peoples view a river as an integral part of its entire ecosystem, encompassing the earth, water, and atmosphere, and teeming with associated wildlife, creating interconnected ribbons and flows between these three spheres (as discussed in Chap. 6), the agroecologist farmer perceives these ribbons of life as conduits for energy to flow from the soil to the trees, then to the crops below, and finally back to the land. This holistic perspective also encompasses the nurturing capacity for all living beings it hosts, whether they are animals, insects, birds, or human beings.

According to Gliessman and Rosemeyer (2010: 6–7; Funes-Monzote, 2010: 223), successful conversion from modern conventional to sustainable agroecological systems involves four levels of transition:

Level 1 Increase the efficiency of conventional practices by reducing waste (energy, water, biomass) and integrating natural elements from the ecosystem, cutting excess chemical use, and leaving/returning vegetation on the site.

[10] There is a growing demand for specialty forest products among gourmet restaurants, retailers, and food enthusiasts (Chamberlain et al., 2020). These products encompass various forest-derived plants that are used to create a wide range of culinary delights, including jams, jellies, salad dressings, marinades, ketchup, or dehydrated ingredients for infusions and cooking.

Level 2 Substitute conventional inputs and practices with alternative practices without impacting the land-use model. Monocultures still exist, but edges of wildflowers are seeded, green manures are planted, and biological pest control is achieved. This system, while more sustainable, retains problems from the agricultural design and layout.

Level 3 Redesign the agroecosystem to function based on a new set of ecological processes and relationships. Implement an ecological design emphasizing DIA: diversity, integration, and autonomy. Transforming the agricultural design at its root prevents old problems, suppressing the need for external inputs to address them. Farm diversification includes reintroducing crop rotations, multiple cropping, and agroforestry systems.

Level 4 Reestablish a more direct connection between those who grow the food and those who consume it, emphasizing an interdependent foodshed. The socioeconomic and cultural context of conversion should be favorable, encouraging a deeper sustainability level of agroecosystems. Consumers support local growers and value access to local and quality food, fostering the development of food citizenship. As these processes become more widespread globally, they contribute to building a new culture and economy of sustainability.

Altieri (1989), often regarded as the father of agroecology, asserts that the primary cause of ecological disorders in conventional agriculture is the monocrop system (In Funes-Monzote, 2010: 223). Various botanical and entomological studies, such as those by Girard et al. (2012) and Koh et al. (2016), identify row crop systems as being responsible for the loss of native pollinator habitats.

This third level of agroecosystem conversion involves reconsidering the layout and size of the farm parcel, adapting it to the crops and tools available. For instance, when Cuba had to decarbonize its agriculture due to a lack of fossil fuel, it reverted to oxen-drawn tools for land cultivation (Funes-Monzote, 2010; Wright, 2009). This shift resulted in improved soil quality, as oxen, compared to tractors, compact the soil less, graze on weeds for natural weed control, and enrich the land with their manure. Reducing parcel size is crucial, moving away from oversized monocrop fields to a more manageable size for a team of oxen and a farmer guiding them through the furrows. This reduction allows for vegetation to grow between parcels, incorporating wildflowers, trees (providing appreciated shade in that climate), and other household activities. Crop rotation and fallow periods contribute to land regeneration.

Envisioning the farmstead as a diversified garden with space for various on-site activities is likely the most sustainable approach to planning its layout. Permaculture and Energy Intensive Agriculture (EIA) (Griffon, 2013, 2017) have demonstrated the advantages of working with small land patches to maximize energy use, including the farmer's energy, solar energy per unit of land for plant growth through photosynthesis, energy emitted by compost degradation, and natural amendments that accelerate microorganism activity in the soil and biodegradation. This approach also maximizes the benefits of green manure to the land, controlling erosion and soil loss, maintaining permanent green cover for natural soil protection and moisture conservation, and promoting proximity between plants for the benefit of species association (Bourguignon & Bourguignon, 2017; Burnett, 2012; Olivier, 2021).

7.15 Toward a Model for Sustainable Farm Design: Notes for Policymakers

The twentieth century witnessed several influential factors that significantly affected changes in land use, subsequently impacting pollinator biodiversity. These factors include:

1. Changes in Farming Techniques, Implements, and Farm Sizes: Agricultural practices evolved with the introduction of modern technologies, machinery, and large-scale farming operations. These changes often led to the alteration of landscapes and a reduction in the diversity of flora and habitats that support pollinators.
2. Crop Homogenization at Farm and Regional Scales: The shift toward monoculture farming, where large expanses of land are devoted to a single crop, reduced the availability of diverse flowering plants that pollinators rely on for food. This resulted in a decline in the variety of nectar and pollen sources, while exposing them to various synthetic pesticides.
3. Urbanization Leading to Construction of Roads and Houses on Agricultural Lands: Rapid urban expansion encroached upon agricultural and forested lands, converting them into urban areas with paved roads, houses, and infrastructure. This transformation disrupted natural wildlife habitats and reduced the availability of flowering plants for pollinators.
4. Industrial Development Competing for Space and Labor with Agriculture: The expansion of industrial sectors often competed with agriculture for land and labor resources. This competition has lead to changes in land use after farm abandonment, such as the conversion of farmland into industrial zones or residential areas.
5. Growth of Leisure Activities and Summer Residences: The development of leisure activities and summer residences in rural areas sometimes resulted in deforestation and land occupation without proper maintenance of native pollen-rich plants. This frequently lead to degraded ecosystems and a decline in pollinator-friendly habitats.

New England forests, while valuable for their ecological role, do not typically provide abundant pollen resources for bees. Only a few tree species, such as red maple, American linden, willow, and yellow birch,[11] are considered melliferous and produce pollen that can support bee populations. Consequently, unmanaged forests may not be the most favorable environments for native bee survival, which may seem counterintuitive in the context of forest conservation efforts. Bees rely on a consistent supply of pollen from flowers throughout the growing season, making the presence of pollen reservoirs crucial for their well-being.

Transitioning from crisis landscapes to ecological transition landscapes involves promoting the development of multifunctional and highly diverse farms and regions. Multifunctionality in land use is essential for supporting wild native bee populations and enhancing their chances of survival.

Here is a simplified list of actions that can be planned at the local and regional scales to initiate the agroecological conversion:

1. Create reconciliation landscapes that integrate humans within their natural environment.
2. Revisit the ecologically diverse farm layout from the early nineteenth century, emphasizing multicrop systems.
3. Draw inspiration from existing agroecological farming practices.
4. Plant butterfly gardens with native pollen-rich plants, including trees, bushes, fruit trees, and flowers that bloom at different times throughout the growing season.
5. Establish flowering meadows and reduce grass-mowing frequency while preserving flowering hedgerows, sideroads, and ditches.
6. Choose natural soil amendments such as ash, straw, compost, vermicompost, animal manure, and green manures.
7. Promote biological weed control techniques, including the use of oxen, goats, or sheep to control weeds and unwanted tree seedlings.
8. Replace invasive plants with native nectar-rich alternatives.
9. Reintroduce diverse functions on farmsteads.
10. Work in harmony with nature rather than against it.
11. Avoid wasting resources, including soil, water, forests, and energy, and aim to close the energy loop on a farmstead.

These actions can help promote sustainable and ecologically friendly practices while supporting native pollinator habitats and biodiversity.[12]

Historical farm designs can serve as inspiration for small organic farmers seeking to cultivate food in harmony with nature, rather than working against it. A farm environment that prioritizes the well-being of pollinators, such as flowering meadows, should be encouraged, fostering the concept of "beescaping" or rewilding the farm. Implementing multiple functions and co-ownership, involving two or more families in the farm's activities, can help diversify income sources while reducing

---

[11] For a complete list, see the Xerces society's native plants for pollinators and beneficial insects guide and for any region of the USA is available online https://xerces.org/publications/plant-lists/native-plants-for-pollinators-and-beneficial-insects-northeast

[12] For tips for a sustainable backyard, see Little (2016) and Tallamy (2012).

the labor burden on a single family and increasing resilience to market fluctuations. Additionally, practicing waste reduction and reusing byproducts or harvest scraps, such as through composting, can lower costs and contribute to closing the energy loop on the farm. These strategies align with agroecological principles and promote sustainable and regenerative farming practices (Olivier, 2021).

Revisiting the pre-1850 multicrop, ecologically diverse, multifunctional, and sustainable small family farm offers an ideal approach for reestablishing a connection with one's foodshed and supplying food to multiple households. This model, characterized by its historical independence from the market and its use of agro-sylvo-pastoral[13] and permaculture techniques, holds great promise for the future of agriculture. It aligns with the goal of conserving biodiversity within and around agricultural fields, making it a compelling and sustainable model for contemporary farming practices.

Drawing inspiration from successful agricultural transitions around the world is crucial for shaping our own sustainable cultivation techniques. For instance, Cuba's shift from conventional to agroecological farming, marked by the reintroduction of animal-drawn machinery, reduced plot sizes, the resurgence of hedgerows, green manure, urban agriculture, and agroforestry[14] offers valuable insights into agricultural decarbonization (Wright, 2009).

Japan's permaculture farms, known as "tekei," which are rooted in traditional farming philosophy, are another noteworthy example. These energy-efficient systems, characterized by practices like using only one cow, avoiding tractors, and intensive mixed-crop cultivation on one acre of land, can provide a model for feeding numerous families sustainably. Masanobu Fukuoka's[15] book, "The One Straw Revolution," encapsulates the fundamental philosophy of permaculture, emphasizing working harmoniously with nature, thoughtful observation over mindless labor, and holistic approaches that consider the multifaceted roles of plants and animals in an ecosystem. This wisdom from diverse farming systems worldwide can guide and inform our own efforts toward sustainable agriculture.

7.16 Conclusion

In conclusion, the decline of farmland biodiversity and the challenges facing pollinators in North America have deep historical roots, exacerbated by modern agricultural practices, urbanization, and changes in societal and economic factors. Since

[13] Agro-silvo-pastoralism is an agroforestry production combining pastoralism (livestock husbandry on pastures).

[14] For example, Cuban pigs feed on royal palm nuts as explained by Ly and Ayala, 2018.

[15] Masanobu Fukuoka is a farmer/philosopher who lives on the island of Shikoku in southern Japan.

the 1940s, monocropping, the elimination of fallow fields, and the use of synthetic pesticides have put significant pressure on on-farm fauna and pollinators.

As agricultural expansion, monoculture systems, and urban development have transformed rural landscape, the number and quality of natural habitats for wild native pollinators have dwindled. The disappearance of mixed-farming biodiverse systems, including agro-silvo-pastoral systems in regions like Maine, has had a profound impact on the availability of pollen, nectar, and nesting grounds for these vital insects.

Historical subsistence farming homesteads once preserved diverse, patchwork-like landscapes that supported wild pollinator populations. However, the industrialization of agriculture has led to homogenized landscapes, posing challenges for pollinator recovery. To address these issues and promote pollinator biodiversity, embracing a genuine agroecological transition within localized agri-food systems is imperative.

This research utilized a comprehensive historical landscape analysis approach, spanning nearly two centuries and focusing on Maine's regions, to assess farmland habitat capacity for sustaining wild native bee populations. The overarching goal was to propose a sustainable farming model that supports pollinator biodiversity at both field and regional scales. By adopting multifunctional, biodiverse, small-scale farming and agroforestry management practices, we can work toward revitalizing and protecting these essential pollinators and the ecosystems they inhabit.

Restoring and valuing Maine's agricultural heritage presents an opportunity to inspire more sustainable agricultural and land-use practices, both at the farm and regional levels. The restoration of old farmland not only combats the decay of heritage farmland but also helps maintain the scenic, humanized, and diversified landscapes that define the state's appeal.

By planting fruit trees and reviving old orchards, we can preserve sought-after ancestral varieties and provide vital resources for pollinators. Managing forests at a regional scale while prioritizing the protection of pollination reservoirs is another important aspect of this restoration effort.

Forest management practices aimed at supporting wild native bees should focus on creating diverse ecosystems with alternating spans of forests, managed forest edges, open flowering meadows, scrublands, oak savannahs, croplands, unmown ditches and roadsides, and unmown stream and riverbanks in a patchwork pattern at the local scale. It is important to consider that a bee's flying distance greatly affects its perception of its surroundings and foraging capacity. When planning a bee-friendly landscape—or beescape—it's crucial to take into account the appropriate scale, which differs from planning for humans.

To translate these findings into actionable steps, it will be necessary to develop and implement public policies that empower citizens and support multifunctional and sustainable family farming throughout Maine. This collaborative approach, already put in action by MOFGA, can contribute to the revitalization of Maine's agricultural traditions and the conservation of its natural resources and biodiversity.

References

Agriculture and Agri-Food Canada. (2014). *Native pollinators and agriculture in Canada Ottawa*. Government of Canada. Retrieved at: https://publications.gc.ca/collections/collection_2014/aac-aafc/A59-12-2014-eng.pdf. Accessed 13 Jan 2024.

Aguilar, S. (ed.) (2009) Areas importantes para la conservacion de las aves en Cuba. La Habana, Ed. Academia. USDA website. The importance of pollinators. Accessed Sept. 1, 2024. https://www.usda.gov/peoples-garden/pollinators

Altieri, M. A. (1989). Agroecology: A new research and development paradigm for world agriculture, Agriculture. *Ecosystems & Environment, 27(1–4), 37–46. Accessed August 26, 2024.* https://doi.org/10.1016/0167-8809(89)90070-4

Banaszak, J. (1992). Strategy for conservation of wild bees in an agricultural landscape. Agriculture, *Ecosystems and Environment, 40, 179–192.*

Berg, P. (2001). "Aux sources du biorégionalisme". *Groupement de recherche et d'études pour la civilisation européenne*. Éléments no 100. Inaccessible online.

Bourguignon, L., & Bourguignon, C. (2017). *Manifeste pour une agriculture durable*. Actes Sud.

Bowler, B., Buggia, L., Johnson, R., & Lloyd McCabe, F. (Eds.). (2012). *Recalling Monmouth's greatest Turf and Farm Era*. Monmouth Museum.

Burnett, G. (2012). *Permaculture: A beginner's guide*. Spiralseed.

Castañeda Abad, Herrera Sorzano, A., & Gónzalez Sousa, R. (2017). Población y organoponía como estrategía de desarrollo local. *Novedades en población. CEDEM, 25*, 43–55.

Celnik, J. (2016). *Les biorégions, nouveaux territoires de la décroissance*. Institut Momentum. Retrieved at: https://www.institutmomentum.org/bioregions-nouveaux-territoires-de-decroissance/. Accessed 13 Jan 2024.

Chamberlain, J. S., Darr, D., & Meinhold, K. (2020). Rediscovering the Contributions of Forests and Trees to Transition Global Food Systems, 11, Accessed August, 28, 2024. https://doi.org/10.3390/f11101098

Colla, S. R. (2016). Status, threats and conservation recommendations for Wild Bumble Bees (Bombus spp.) in Ontario, Canada: A review for policymakers and practitioners. *Natural Areas Journal, 36*(4), 412–426. Retrieved at: https://www.jstor.org/stable/90009234. Accessed 13 Jan 2024.

Colston Burrell, C. (2006). *Native alternatives to invasive plants*. Brooklyn Botanical Garden.

Curtis, A. (2019). Maine is losing farms and farmland, but hope is not lost for agriculture. *Bangor Daily News*. Retrieved at: https://www.bangordailynews.com/2019/05/05/homestead/maine-is-losing-farms-and-farmland-but-hope-is-not-lost-for-agriculture/. Accessed 13 Jan 2024.

Daum, T. (2023). Mechanization and sustainable agri-food system transformation in the Global South. A review. *Agronomy for Sustainable Development, 43*(16). Retrieved at: https://doi.org/10.1007/s13593-023-00868-x. Accessed 13 Jan 2024.

Du Clos, B., Drummond, F. A., & Loftin, C. S. (2022). Effects of an early mass-flowering crop on wild bee communities and traits in power line corridors vary with blooming plants and landscape context. *Landscape Ecology*. Retrieved at: https://doi.org/10.1007/s10980-022-01495-9. Accessed 13 Jan 2024.

Edwards, K. (2017). Western Kennebec County towns look for ways to keep farmers on farms. *Central Maine*. Retrieved at: https://www.centralmaine.com/2017/10/24/western-kennebec-county-towns-look-for-ways-to-keep-farmers-on-the-farm/. Accessed 13 Jan 2024.

Everaars, J., Settele, J., & Dormann, C. F. (2018). Fragmentation of nest and foraging habitat affects time budgets of solitary bees, their fitness and pollination services, depending on traits: Results from an individual-based model. *PLoS ONE, 13(2), e0188269. Accessed August 28, 2024.* https://doi.org/10.1371/journal.pone.0188269

FAO website Evaluating agri-food systems. Accessed Sept. 1, 2024. https://www.fao.org/evaluation/highlights/agri-foodsystems/en#:~:text=A%20sustainable%20food%20system%20is,future%20generation%20is%20not%20compromised

Falk, C. G. (2012). *Barns of New York: Rural Architecture of the Empire State*. Cornell University Press.

Fater, L. (2020). The lost art of growing blueberries with fire. *Atlas Obscura*. Retrieved at: https://www.atlasobscura.com/articles/why-do-people-burn-blueberry-fields. Accessed 13 Jan 2024.

Funes-Monzote, F. R. (2010). Cuba: A national-level experiment in conversion. In S. R. Gliessman & M. Rosemeyer (Eds.), *The conversion to sustainable agriculture. Principles, processes, and practices* (pp. 205–238). CRC Press.

Gamboa Costa, A., Herrera Sorzano, A., & Gónzalez Sousa, R. (2013). Soberanía y seguridad alimentaria en Cuba: politicas públicas necessarias para reducir la dependencia alimentaria. *Agrisost, 19*(3), 1–14. Retrieved at: http://www.agroecologynetwork.org/uploads/4/9/2/9/49299363/gamboa_et_al_2013_seguridad_y_soberania_alimentria_en_cuba.pdf. Accessed 13 Jan 2024.

Girard, M., Chagnon, M., & Fournier, V. (2012). Pollen diversity collected by honeybees in the vicinity of Vaccinium spp. crops and its importance for colony development. Pollination biology research in Canada: Perspectives on a mutualism at different scales. *Botany, 90*(7), 545–555. Retrieved at: https://www.semanticscholar.org/paper/Pollen-diversity-collected-by-honey-bees-in-the-of-Girard-Chagnon/fbc94ad315aa98fb5059b5adf345e120d8ad9c77. Accessed 13 Jan 2024.

Gliessman, S. R., & Rosemeyer, M. (2010). *The conversion to sustainable agriculture. Principles, processes, and practices*. CRC Press.

Gravel, N. (2004). "Faire plus avec moins": patrimoine domestique et survie au Mexique rural, 1990–2002. *Les Cahiers de géographie du Québec, 48*(134), 156–172. Retrieved at: https://www.erudit.org/fr/revues/cgq/2004-v48-n134-cgq990/011679ar/. Accessed 13 Jan 2024.

Gravel, N., & Mestiri, R. (2015). Le jardin communautaire comme pôle de gouvernance environnementale en milieu périurbain havanais, Cuba. Presentation at the 83rd congress of ACFAS, Rimouski, Quebec.

Gravel, N. (2017). *Cities as safe havens for pollinators?* Paper presented at the annual conference of the canadian association of geographers (ACG). York University, Toronto, May 29th–June 2nd.

Griffon, M. (2013). *Pour des agricultures écologiquement intensives*. Nouvelles éditions de l'Aube.

Griffon, M. (2017). *Écologie intensive : la nature comme inspiration pour l'agriculture et la société. Paris: Buchet-Chastel.*

Hardy, J. W. (2008). *Settlement and Abandonment on Tatnic Hill, An Eclectic History of Wells, Maine 1600–1900*. Back Channel Press.

Herrera Sorzano, A. (2018). La agricultura en Cuba y las reformas agrarias. In L. F. Rincón & R. Kretschmer (Eds.), *B. Mançano Fernandes* (pp. 141–166). La actualidad de la reforma agrarian en America Latina y el Caribe, CLACSO.

Hornsby, S. J., & Judd, R. W. (2015). *The historical Atlas of Maine*. The University of Maine Press.

Horwitz, R. P. (1978). *Anthropology toward history. Culture and work in a 19th-century Maine town*. Wesleyan University Press.

Hunter, D., & Lightner, J. (2016). *Mason Bee revolution. How the hardest working Bee can save the world one backyard at a time*. Skipstone.

Jepsen, S., Evans, E., Thorp, R., Hatfield, R., & Hoffmann Black, S. (2013). *Petition to list the rusty patched bumblebee Bombus affinis (Cresson) 1863 as an endangered species under the US Endangered Species Act*. Retrieved at: https://www.researchgate.net/publication/270162303_Petition_to_list_the_rusty_patched_bumble_bee_Bombus_affinis_Cresson_1863_as_an_Endangered_Species_under_the_US_Endangered_Species_Act/figures?lo=1. Accessed 13 Jan 2024.

Kloppenburg, J. J., Hendrikson, J., & Stevenson, G. W. (1996). Coming in to the foodshed. *Agriculture and Human Values, 13, 33–42. Retrieved at:* https://link.springer.com/article/10.1007/BF01538225 Accessed January 18, 2024.

Koh, I., Lonsdorf, E. V., Williams, N. M., Brittain, C., Isaacs, R., Gibbs, J., & Ricketts, T. H. (2016). Modeling the status, trends and impacts of wild bee abundance in the United States. *Proceedings of the National Academy of Sciences, 1*(113), 140–145. https://doi.org/10.1073/pnas.1517685113. Accessed 13 Jan 2024.

Leopold, D. J. (2005). *Native plants of the Northeast. A guide for gardening and conservation.* Timber Press.

Little, S. (2016). *Introduction to organic lawns and yards. Plus a checklist for an eco-friendly property.* Northeast Organic Farming Association.

Luthra, A., Cunningham, K., Fraser, A. M., Pandey, A., Rana, S., & Singh, V. (2023). Ecological livelihoods of farmers and pollinators in the Himalayas: Doing critical physical geography using citizen science. *Canadian Geographies, 67*, 35–51. https://doi-org.acces.bibl.ulaval.ca/10.1111/cag.12799

Ly, J., & Ayala, L. (2018). Pig breeding with royal palm nuts and other palm byproducts. *Cuban Journal of Agricultural Sciences, 52*(4), 375–382. Retrieved at: https://www.researchgate.net/publication/343572634_Pig_breeding_with_royal_palm_nuts_and_other_palm_by-products_Short_review/citations#fullTextFileContent. Accessed 13 Jan 2024.

Mader, E., et al. (2011). *Attracting native pollinators – Protecting North America's bees and butterflies.* Story Publishing.

Maine State Grange. (1896). *Journal of Proceedings of the Fifty-Fifth Annual Session of the Maine State Grange*, Augusta City Hall, December 11–13th.

Maine State Grange. (1928). *Journal of Proceedings of the Fifty-Fifth Annual Session of the Maine State Grange*, Augusta City Hall, December 11–13th.

Maine State Grange. (1930). *Journal of Proceedings of the Fifty-Fifth Annual Session of the Maine State Grange*, Augusta City Hall, December 11–13th.

Martins, K. T., Gonzalez, A., & Lechowicz, M. J. (2015). Pollination services are mediated by bee functional diversity and landscape context. *Agriculture, Ecosystems & Environment, 200*, 12–20. Retrieved at: https://www.sciencedirect.com/science/article/abs/pii/S0167880914004812. Accessed 13 Jan 2024.

Marris, E. (2011). *Rambunctious Garden: Saving Nature in a Post-Wild World. New York: Bloomsbury.*

McFrederick, Q. S., & LeBuhn, G. (2006). Are urban parks refuges for bumble bees Bombus spp. (Hymenoptera: Apidae)? *Biological conservation, 129*(3), 372–382. Retrieved at: https://melittology.files.wordpress.com/2013/03/biol-conserv-2006-mcfrederick.pdf. Accessed 13 Jan 2024.

McGreevy, S.R., Rupprecht, C. D.D., Niles, D. et al. (2022). Sustainable agrifood systems for a post-growth world. *Nat Sustain, 5, 1011–1017. Accessed Sept. 1, 2024.* https://doi.org/10.1038/s41893-022-00933-5

McMurry, S. (1995). *Transforming rural life. Dairying families and agricultural change, 1820–1885.* John Hopkins University Press.

Melvin, M. (2022). Pollinating Orchards by Hand. Lessons from Sichuan, China. Foodunfolded Accessed August, 28, 2024 https://www.foodunfolded.com/article/pollinating-orchards-byhand-lessons-from-sichuan-china

Mortillaro, N. (2022). Nearly half of the world's birds are on the decline, which experts say is a serious threat to ecosystems. *CBC News*. Retrieved at: https://www.cbc.ca/news/science/birds-declines-1.6610220. Accessed 13 Jan 2024.

Nalepa, R., & Colla, S. R. (2023). Toward a wild pollinator strategy for Canada: Expert-recommended solutions and policy levers. *FACETS, 8*, 1–18. Retrieved at: https://doi.org/10.1139/facets-2022-0204. Accessed 13 Jan 2024.

Nelson, F. A. (1945). *Lest we forget.* Portland Publishing.

Normandin, É., Fournier, V., & Rivard, N. (2015). Les abeilles sauvages: une étude de leur diversité. Parc national des Iles-de-Boucherville. *Bulletin de Conservation*, 20–23. Retrieved at: https://www.sepaq.com/dotAsset/7062d8d3-1476-4fb9-8ccf-fe230257ec81.pdf. Accessed 13 Jan 2024.

Olivier, A. (2021). *La révolution agroécologique. Nourrir tous les humains sans détruire la planète.* Écosociété.

Olmstead, A. (1975). The mechanization of reaping and mowing in American agriculture, 1833–1870. *The Journal of Economic History, 35*(2), 327–352. Retrieved at: https://doi.org/10.1017/S0022050700075082. Accessed 13 Jan 2024.

Padrón Padrón, W. (2015). *Mejorar la capacidad productiva del suelo en la Finca "El Pedregal". Incremento de su biodiversidad. Qué se ha hecho?* Paper presented at the Living Soil Symposium, Journées sur les sols vivants, Quebec City.
Renault, C. (2017). À Cuba les abeilles sont en très bonne santé car il n'y a pas de pesticides. *Sain et naturel*. Retrieved at: https://sain-et-naturel.ouest-france.fr/cuba-abeilles-pas-de-pesticides.html. Accessed 13 Jan 2024.
Reuters. (2016). Cuban honey is a sweet success for Cuba as other bee populations suffer. *The Guardian*. Retrieved at: https://www.theguardian.com/world/2016/feb/09/organic-honey-is-a-sweet-success-for-cuba-as-other-bee-populations-suffer#:~:text=Organic%20honey%20is%20a%20sweet%20success%20for%20Cuba%20as%20other%20bee%20populations%20suffer,-This%20article%20is&text=Long%20known%20for%20its%20cigars,to%20declining%20bee%20populations%20elsewhere. Accessed 13 Jan 2024.
Rosenzweig, M. (2003) Reconciliation ecology and the future of species diversity. Oryx. 37(2):194–205. Accessed Sept. 1, 2024. https://doi.org/10.1017/S0030605303000371
Sale, K. (1996). Principles of Bioregionalism. In J. Mander & E. Goldsmith (Eds.), *The case against the global economy and for a turn toward the local* (pp. 471–484).
Sanford, R. (2015). *Reading rural landscapes. A field guide to New England's past*. Tilbury House.
Silber, T. (1988). *A small farm in Maine*. Houghton Mifflin.
Stuppia, P. (2022). Reinhabiting Earth. 50 years of Bioregional Utopias. *Développement durable et territoires*. Special issue on Territories during the Pandemic/Experimentations of Ecological Transitions, *13*(2). Retrieved at: https://doi.org/10.4000/developpementdurable.21572. Accessed 13 Jan 2024.
Tallamy, D. W. (2012, reed. 2018). *Bringing nature home. How you can sustain wildlife with native plants* (Updated and expanded). Timber Press.
The Xerces Society. (2011). *Attracting native pollinators. Protecting North America's bees and butterflies*. Story Publishing, 371 p.
Thoreau, H. D. (1950). *The Maine Woods*. Arranged by D. C. Lunt. The Excursions of Henry David Thoreau through the Woods of Maine in the Years 1846, 1853 and 1857. W.W. Norton.
Tompkins, P., & Bird, C. (1989). *Secrets of the soil*. Harper & Row.
van Vierssen Trip, N., MacPhail, V. J., Colla, S. R., & Olivastri, B. (2020). Examining the public's awareness of bee (Hymenoptera: Apoidae: Anthophila) conservation in Canada. Conservation Science and Practice, vol. 2 (12). Accessed Sept. 1. 2024. https://doi.org/10.1111/csp2.293
Venturini, E. M., Drummond, F. A., Hoshide, A. K., Dibble, A. C., & Stack, L. B. (2017). Pollination reservoirs for wild bee habitat enhancement in cropping systems: A review. *Agroecology and Sustainable Food Systems, 41*(2), 101–142. Retrieved at: https://doi.org/10.1093/jee/tow285. Accessed 13 Jan 2024.
Ward, T., & Adamson, N. (2015). Incorporating pollinator habitat into silvopasture. USDA National Agroforestry Center. *Inside Agroforestry, 23*(2), 4–7. Retrieved at: https://www.fs.usda.gov/nac/assets/documents/insideagroforestry/IA_vol23issue2_pollinators.pdf. Accessed 13 Jan 2024.
Weissenstein, M. (2016). Cuba reports remarkable success in containing Zika virus. *CTV News*. https://www.ctvnews.ca/health/cuba-reports-remarkable-success-in-containing-zika-virus-1.3055359. Accessed 27 May 2024.
Wicks, G. A., & Simka, D. E. (1973). Chemical fallow in a winter wheat-fallow rotation. *Weed Science, 21*(2), 97–102. Retrieved at: https://www.jstor.org/stable/4042053?typeAccessWorkflow=login. Accessed 13 Jan 2024.
Wright, J. (2009). *Sustainable agriculture and food security in an era of oil scarcity. Lessons from Cuba*. Routledge.
Yang, Y., Hobbie, S. E., Hernandez, R. R., Fargione, J., Grodsky, S. M., Tilman, D., Zhu, Y.-G., Luo, Y., Smith, T. M., Jungers, J. M., Yang, M., & Chen, W.-Q. (2020). Restoring abandoned farmland to mitigate climate change on a full earth. *One Earth, 3*(2), 176–186. Retrieved at: https://doi.org/10.1016/j.oneear.2020.07.019. Accessed 13 Jan 2024.

Websites

Chicago Botanic Garden website. *Witch hazel.* https://www.chicagobotanic.org/plantinfo/witch_hazel. Accessed 13 Jan 2024.
Food Ethics Council website. Accessed Sept. 1, 2024. https://www.foodethicscouncil.org/programme/food-citizenship/
Food and Agriculture Organization (FAO). (2016). *Comment pouvons-nous protéger les pollinisateurs et souligner leur importance dans les pratiques environnementales et agricoles?* http://www.fao.org/publications/card/fr/c/b2f54b5b-2824-49de-bbdb-624089b05e29/. Accessed 13 Jan 2024.
FAO website Evaluating agri-food systems. Accessed Sept. 1, 2024. https://www.fao.org/evaluation/highlights/agri-foodsystems/en#:~:text=A%20sustainable%20food%20system%20is,future%20generation%20is%20not%20compromised
Maine- An Encyclopedia. https://maineanencyclopedia.com/grange-the/. Accessed 13 Jan 2024.
Maine Historical Society. (2000). *Maine history online. Farming in Maine.* Maine Memory Network Project. https://www.mainememory.net/sitebuilder/site/161/slideshow/192/display?format=list&slide_num=1. Accessed 13 Jan 2024.
Minnesota Landscape Arboretum. (2021). *Rusty patched Bumble bees at the Arb.* https://arboretumnaturenotes.wordpress.com/2021/07/29/rusty-patched-bumble-bees-at-the-arb/. Accessed 13 Jan 2024.
Organoponico Vivero Alamar 2018 video. Accessed Sept. 1. 2024. https://www.youtube.com/watch?v=K1fh3CSHWNE
The Canadian Press. (2012). *Bird species declining in Canada report says.* https://www.cbc.ca/news/canada/bird-species-declining-in-canada-report-says-1.1149294. Accessed 13 Jan 2024.
The Xerces Society for Invertebrate Conservation. (2023). *Native plants for pollinators and beneficial insects guide: Northeast.* https://xerces.org/publications/plant-lists/native-plants-for-pollinators-and-beneficial-insects-northeast. Accessed 13 Jan 2024.
U.S. Department of Agriculture (USDA). *Wildlife and pollinators.* U.S. Forest Service. https://www.fs.usda.gov/nac/topics/wildlife-pollinators.php. Accessed 13 Jan 2024.
U.S. Fish and Wildlife Service website. (2017). *Rusty patch bumble bee Bombus affinis factsheet.* https://www.fws.gov/sites/default/files/documents/508_RPBB-factsheet.pdf. Accessed 13 Jan 2024.
Union des producteurs agricoles (UPA). *Le goglu des prés.* Québec, Canada. https://www.upa.qc.ca/fileadmin/monteregie/Document_en_ligne/2_Goglu_des_pres.pdf. Accessed 27 May 2024.
University of Maine, BeeMapper Group. https://umaine.edu/beemapper/development/. Accessed 13 Jan 2024.

Chapter 8
Seeds of Change: Urban and Peri-Urban Agriculture Dynamics in Brazil

This chapter is dedicated to the loving memory of the urban agriculture activist, Sergio Shigeeda, who passed away in 2022.

Abstract In Brazil, urban and peri-urban agriculture (UPA) faces challenges to thrive despite being part of anti-hunger initiatives. This chapter explores UPA actors' attempts to influence municipal policies addressing local food needs and fostering green cities. Urban agriculture offers numerous benefits, such as improving diet, fostering social connections, providing leisure activities, and greening cities. Decision-makers should recognize its potential to combat poverty and malnutrition. Using the lens of Social Regulation Theory, this chapter examines garden coordinators' efforts and advocacy for legal frameworks, recognition, and urban planning policies to facilitate and democratize access to UPA. The study of three metropolises in Brazil allowed for actor-mapping, learning from UPA public policy-making, and making comparisons. The institutionalization of UPA in Salvador de Bahia and Florianópolis provides a model, emphasizing the role of the Health Ministry. Beyond "guerrilla gardening" present in well-to-do neighborhoods of São Paulo, the future of UPA in that metropolitan region may lie at its periphery. These areas can foster stable and productive fresh food production in proximity, contributing to a large array of family incomes. Developing public policies aligned with gardeners' preferences can help institutionalize UPA, making it a cornerstone of urban sustainability and community food resilience in the face of climate change.

Keywords Actor-mapping · Anti-hunger initiatives · Brazil · Florianópolis · Food security · Fresh food production · Green cities · Guerrilla gardening · Health Ministry · Institutionalization · Metropolis · Multifunctionality · Peri-urban areas · Poverty reduction · Public policies · Salvador de Bahia · São Paulo · Social Regulation Theory (SRT) · Urban and peri-urban agriculture (UPA)

N. Gravel, *Tending Nature*, https://doi.org/10.1007/978-3-030-47062-3_8

8.1 Introduction[1]

The persistence of green spaces where people can participate in activities related to gardening in urban and peri-urban areas is not uncommon but may be endangered by urban expansion and the densification of cities. The multifunctionality of peri-urban areas, characterized by lower human habitat density and diversified land uses, often includes industrial activities, ploughed fields, abandoned sites, green belt land protection, traditional farmhouses, and sometimes the last remnants of forest vegetation until the city expands further. What will happen to the last agricultural lands that have endured at the periphery of major cities and feed their inhabitants? Are we taking any measures to ensure the preservation of this multifunctionality and the diversity of ways of living in peri-urban areas, or is there a trend toward homogenization that goes unnoticed while it transforms natural spaces into built environments? What role could urban and peri-urban agriculture (UPA) play in providing and protecting green areas in both the inner and outer city?

This chapter aims to contribute to the reflection on public policies that could better support the development of sustainable food systems both within and outside of metropolises. Special attention will be given to the groups of actors who are considered changemakers in the context of municipal gardening activities. Reynaud's Social Regulation Theory (SRT) has been chosen to appraise the tensions between urban agriculture (UA) actors and the municipal decisional system and to assess the outcomes in terms of urban garden governance and the future (Reynaud, 1997, 1999; De Terssac, 2012).

Studying the UPA movement in Brazil can be approached from various perspectives, such as the food security perspective, which is relatively well documented in developing countries for its income-generating and food production capacity (Mougeot, 2005a; Hovorka, 2005; De Bon et al., 2010; Mok et al., 2013; Gravel & Mestiri, 2015). However, this perspective does not encompass the entire reality of UPA in Brazil. In the city of São Paulo, far from the *favelas* (slums), urban gardeners create projects and fight for the right to the city by producing healthy food locally, educating the neighboring population about gardening and eating locally, and sharing and exchanging knowledge. These urban gardening guerrilleros, as they are sometimes referred to, follow in the footsteps of the founders of the UPA movement in the United States in the 1960s and 1970s (Visoni & Nagib, 2019), which was motivated by the right to cultivate land in urban areas and preserve the diversity of urban land uses. Often confronted with policymakers who were not always receptive to their demands, these groups decided to focus on claiming green spaces to be used for citizen initiatives.

[1] A first version of this research has been published as Gravel, N. and J. Lemieux (2022) « L'agriculture urbaine et périurbaine au Brésil, une dynamique de lutte pour l'appropriation de l'espace public par les citoyens» in *Les Cahiers de Géographie du Québec*, 27 (187). https://doi.org/10.7202/1112468ar

In view of this, adopting the perspective of Reynaud's Social Regulation Theory (SRT) makes sense to study the dynamics of demands, confrontations, and negotiations that actors from civil society experience when they try to influence the management and regulation of urban green areas (Gravel, 2017). First, it is important to identify the groups of actors involved and the support they receive or the obstacles they face in trying to secure collective or community gardening spaces within their metropolitan regions. The stories of their struggles often illustrate the evolution and implementation of new regulations after years of "fighting the system" in cases where municipal authorities are not supportive from the beginning.

This chapter is based on a multisite investigation conducted in Brazil between 2014 and 2016 (and follow-up questions in 2019 and 2020 at a distance). It presents case studies of three Brazilian metropolises (São Paulo, Salvador, and Florianópolis) that progressed at different rhythms in the development of their UPA policies. Their study will help shed light on current UPA planning and management problems and understand the determining factors in the emergence and continuity of community UPA initiatives.

8.2 What's at Stake in Food Production in Global South Cities?

While home gardens, livestock, and market garden production areas were already part of the urban fabric since the appearance of human settlements, the twentieth century was marked by a separation of the productive functions from consumption sites, mainly driven by the monetary value of lands located at the heart of cities. That trend toward the growing separation of cities and food production would have continued were it not for the persistence of some engaged citizens who fought to maintain a diverse urban landscape comprising mixed-use lands, green spaces, and blue components and promoted its benefits. This chapter explores the emergence, coexistence, and the perceived future of various forms of urban and peri-urban agriculture (UPA) in Brazilian cities, which are shaped by varying degrees of politicization.

In the context of the Global South, "the poorest of urban citizens have to cultivate on risky, poorly located lands that are sometimes contaminated to produce their food," which is one of the consequences of what is known as the "urbanization of poverty" (Arlego Paraguassu, 2013: 54). The urbanization of poverty emerged notably in the 1940s as rural-to-urban migrations intensified, primarily propelled by farm mechanization. As agricultural practices modernized, many unemployed rural inhabitants sought better opportunities in urban areas, leading to significant demographic shifts. However, these migrants often carried their poverty with them, contributing to the proliferation of informal settlements on the outskirts of cities. This phenomenon was particularly pronounced in Brazil, where the growth of favelas mirrored the influx of rural migrants seeking livelihoods in urban environments

(Gravel, 2009; Rivière D'Arc, 2006; Bricas & Conaré, 2019). Successive waves of migrants have populated peri-urban areas, significantly increasing the number of people living in precarious conditions and facing food insecurity issues. This influx has contributed to the escalation of violence and poverty in South American metropolises, as noted by Perlman (2010).

By 1960, Brazil's urban population surpassed its rural population, accounting for 55.92% of the total population (Arlego Paraguassu, 2013: 110). According to the World Bank, Brazil's urbanization rate reached 87.56% by 2022 (Trading Economics, 2022). Over recent decades, Brazilian cities have welcomed 30 million new inhabitants due to internal migration, settling in disorganized patterns on the outskirts and in conurbation areas where the urban fabric is continuous, encompassing numerous cities (Idem). The São Paulo metropolitan area alone comprises 39 cities!

Urban and peri-urban agriculture represents an opportunity for reintroducing food production in cities and reversing the trend toward the separation of the food consumption and production functions. Developing localized sustainable food system represents a promising path toward helping cities regain their alimentary function and relocate some of their livestock and market fresh garden produce near their citizens while reducing their greenhouse gas emissions associated with transportation. The Milan urban food policy pact (signed by 180 cities in 2015, it had over 280 signatory cities in 2024) encouraged cities to reclaim their alimentary function by creating meeting spaces for urban consumers and producers, promoting not only foodstuff trade but also the consolidation of social bonds between the actors in the food value chain. By emphasizing the need to develop sustainable localized food systems, the signatories agreed to exchange best practices in a city-to-city cooperation, and their representatives attend annual global events; rewards are given to cities for their innovative approaches in food system transformation (Bricas & Conaré, 2019: 8; Milan Urban Food Policy Pact website).

Having access to a diverse and healthy diet is one of the ways through which individuals can increase their capacities and human capital, which, in turn, translates into a better potential for the economic, social, cultural, environmental, and political development (according to Bebbington's five capitals model) of their households and population groups, thanks to the interchangeability of capitals (Bebbington, 2004). That is, access to good, clean, and culturally appropriate food on a regular basis act as a lever for people and groups to undertake other activities that contribute to their own development or that of their group. "The UNPD [United Nations Development Program] identifies UA as one of the existing strategies to achieve human development in the cities" (Arlego Paraguassu, 2013: 43). During the Second United Nations Conference on Human Settlements (Habitat II), that question took a central place in the debates to the point where it led to the redaction of a document: *Urban Agriculture: Food, Jobs, and Sustainable Cities* (1996).

According to Mougeot (2005b), there is a lack of studies regarding the understanding of how UA contributes to the development of the individuals of the poorest households in the global South. To address this lack, this chapter will focus on a sample of Brazilian cities that were able to build on the national agroecology movement—considered a pioneer in the world—and on an early regulatory framework that allowed them to concentrate efforts and funds at the national and regional scales to fight against hunger.

8.3 Urban Agriculture: Definitions and Safety Considerations

Intraurban and peri-urban agriculture is polymorphous and adapts to the different possibilities presented by the milieus where they occur. This research will consider only the form of community gardens, which results from the free association of urban farmers who come together as members of an organization or respond to an invitation made by municipal authorities, often led by a chief urban gardener (coordinator hired by the city or not). Other types of intraurban and peri-urban agriculture include institutional gardens (schools, hospitals, prisons, etc.), home gardens, nurseries, and rooftop gardens, but they will be omitted for the purpose of this research. Whether inside of the urban core or at the periphery, although the available space may differ, the forms are sensibly the same. When trying to understand the differences between urban and rural agriculture, Mougeot (2000: 10) defined urban and peri-urban agriculture, not so much by its location but by "its integration into the urban economic and ecological system":

> Urban agriculture is an industry located within (intraurban) or on the fringe (peri-urban) of a town, city or metropolis and grows or raises, processes and distributes a diversity of food and nonfood products, (re)uses largely human and material resources, products and services found in and around that urban area, and in turn supplies human and material resources, products and services largely to that urban area.

According to Mougeot (Idem:17), most urban farmers are men and women with a low income who cultivate for home consumption on small lots that they do not own, with very little protection. Another trait that characterizes UPA is the insecurity of land holding, as these urban farmers do not have land tenure and are subject to any change in the title of the property or urban development project. This vulnerability can be perceived since they are reluctant to invest in any fixed and durable structure. Policymaking should tackle this problem of lack of permanence to guarantee their presence and secure their land since UPA is an important part of urban sustainability (FAO, 2007).

Among the benefits derived from UPA, we may consider its contribution to household nutrition, the productive occupation of open and vacant lands (instead of leaving them to gangs, delinquents, or as improvised dumping sites) and the cleaning and safe reuse of organic waste (Mougeot, 2000, 2005a). We may add to this list the creation of ecologically diversified living environments, the sense of bonding and belonging to a community, the procurement of a stress-relieving pause favorable to urbanites' mental health, the possibility to renew or maintain a link with nature, and the creation of knowledge trading and educational sites. Far from saying that households enjoy food autonomy out of their gardening activities, UPA contributes to adding freshness and variety to urban household members' diets and can be an important component of their nutrition.

According to Carlos Petrini (2007), founder of Slow Food and initiator of programs such as Slow Food's "A Thousand Gardens in Africa," whose mission is to provide populations with healthy, quality, locally produced food, the quality of life depends on the capacity to have fresh products on a regular basis (Slow Food, 2014). Producing the food we eat or buying it directly from the farmer guarantees its freshness, safety, and conformity to the local eating culture. These precepts are in line with the concept of food security, whether qualitative or quantitative, and include meeting populations' nutritional needs (Petrini, 2007; FAO, 2010).

Nevertheless, the potential public health implications of UA should not be neglected, especially regarding livestock production and the use of animal waste as compost (contamination risks) (Mougeot, 2000), but the literature shows that benefits far exceed potential risks (Aubry & Consalès, 2014). Regarding industrial risks, it is worth mentioning that the use of agrochemicals exists in shared gardens but remains rare. Peri-urban productions have less access to organic fertilizers made from urban waste than intraurban productions, but they tend to use animal waste more frequently to enrich soils.

Cultivating on contaminated soils and the presence of air contaminants in cities (near highways, for instance) is also a topic that is worthy of attention (Baudelet, 2015; Taguchi & Santini, 2019: 16; FAO, 2007). A study conducted in three urban gardens located in São Paulo showed that the lead and cadmium concentrations in leafy vegetables exceeded the limit values for consumption in gardens located next to main highways (Amato-Lourenco et al., 2016). The contamination risks decrease when the garden is located further away from major avenues and with the presence of vertical obstacles (e.g., trees) (Idem). Some soils contaminated by creosote-treated timber next to outdoor furniture and structures (e.g., play structures for children in parks, wooden decks, or railroads) cannot be used for agriculture for several years (at least 20). In this type of risky environment, using raised garden beds with contaminant-free soil may be an alternative. Using pure and, ideally, chlorine-free water is also strongly recommended.

8.4 Environmental Benefits of Urban and Peri-Urban Agriculture

The FAO report (2007) acknowledges that the environmental benefits of UPA can be compared to those of urban parks, and its multifunctional nature makes it even more essential to urban populations. If we focus only on the ecological functions of UPA, we can list the following elements: the greening of urban spaces (green cities) (Cockrall-King, 2016), lowering temperatures under plant cover (by reducing heat islands) (Jonsson, 2004), a reduction in greenhouse gas emissions associated with food transportation (Kulak et al., 2013), and biodiversity regulation through insect pollination, given that gardens provide pollinators with food (pollen) and clean water to drink (bees drink water!) and a natural habitat where they can nest (Cockrall-King, 2016: 111; Dumas, 2018; MEA, 2005; Hunter & Lightner, 2016; Braga Moraes et al., 2018).

Since 2010, authorities have acknowledged the importance of composting to fertilize soils for food production in urban and peri-urban areas (Nagib, 2016). Producing compost with organic waste from households and urban food fairs is relatively common in Brazil and is often regulated. In São Paulo, collection and processing sites have been established to make use of vegetable waste produced during food fairs, and the compost is used in public park plantations and community/collective gardens. The city of Florianópolis, in southern Brazil, has innovated with a first law promoting composting to create fertilizers to be used by urban gardens (Di Fiore et al., 2022).

The implementation of UPA projects on vacant lots and the subsequent revegetation processes enhance the urban landscape and make it possible to find a balance in the composition of the urban fabric between built areas and green spaces. They can even contribute to containing property development, as reported by Aubry et al. (2012) in Antananarivo, Madagascar. These authors also noted that UA makes it possible to clean up the city from organic wastes that would otherwise end up in backyards or landfills (idem). The expanded and, sometimes, continuous green infrastructures—taking the form of urban corridors—allow for the movements and survival of terrestrial and bird species and help increase urban and peri-urban biodiversity. It fosters a better balance between the several natural functions of urban areas (a better water cycle thanks to the presence of porous material and roots in the ground (sometimes following the removal of permeable surfaces), air purification thanks to tree plantation, better water quality due to natural, pesticide-free crops, hillside erosion control using tree plantation, and retaining walls and support for microbial life in soil) (Cockrall-King, 2016; Duchemin, 2013, Wegmuller and Duchemin, 2010).

In this context, our suggestion is to view urban gardens as elements that contribute to the building of sustainable cities in support of the food and nutrition security of the most vulnerable families (without excluding others). Shared gardens make it possible to tackle the challenges of inequitable access to green spaces in urban areas, as they were highlighted by the MEA (2005). The authors of the MEA claim that "Urbanization is not inherently bad for ecosystems," although it modifies the soil surface and often results in deforestation. The question is as follows: How can we reduce the negative impacts of urbanization on ecosystems? The answer might originate in part from the same authors (Idem: 35):

> Ecosystems in and around urban areas can provide a high level of biodiversity, food production, water services, comfort amenities, cultural values and so on if well managed. When the loss of ecosystem services due to urban activities is systematically addressed, these losses can be greatly reduced.

In view of this overview of UPA, the questions that we want to answer in this chapter are the following: (1) What forms does UPA take in Brazil? (2) How can UPA contribute to rewilding cities? (3) What latitude do UPA actors have to carry out their projects and influence public policies?[2]

8.5 Methodology

Data collection for this research lasted over 2 years and focused on the metropolis of São Paulo, Florianópolis (state capital of Santa Catarina, southern Brazil), and the city of Salvador de Bahia (capital of the state of Bahia, northeastern Brazil) (see Fig. 8.1). Data collection took place during several stays between 2014 and 2016, one as guest researcher at the São Paulo Forest Institute (*Instituto Florestal de São Paulo*), which lasted 5 months in 2016. Data collection at the other two cities was organized as two 1-week fieldtrips departing from the São Paulo research base. All of these stays made it possible for the author to accompany graduate students and interns of the Canada–Brazil joint research program, as well as to conduct 30 interviews with urban garden organizers, urban gardeners, NGO representatives, professionals, and city employees in all three metropolitan areas. Field work included participation in the São Paulo Green Belt Biosphere Reserve program of activities, visiting community gardens and attending UA association meetings in all three cities.

The choice of these agglomerations was made based on contacts established in universities from the network on public policies in Latin America, of which I am a member. Additionally, the different degrees of development of UPA initiatives motivated this choice to obtain the widest possible panorama of the different applications and experiences of UPA in Brazil. Importantly, the city of São Paulo stood out by establishing the first laws governing the creation of shared gardens as early as 2004 (de Lima Caldas & Jayo, 2019); Salvador de Bahia followed suit in 2008 with a different governance approach incorporated into the Ministry of Health, which quickly gave conclusive results; and, finally, Florianópolis experienced a later implementation that combined both citizen-based science and an institutional framework, with results that are variable due to the local and national political context.

The interviews conducted with the urban garden coordinators[3] adopted the following guidelines:

- Coordinator's profile (life trajectory, motivations, achievements)
- History of the urban garden, funding, and normative framework

[2] In this paper, public policy is defined as "a document written by government actors presenting their perspective on a matter that could generate a public action and, incidentally, the legal, technical, practical and operational aspects of that action" (Turgeon & Savard, 2012: 1) (Translated from French by the author).

[3] Also referred to here as UPA leaders and activists.

Fig. 8.1 Study sites locations in Brazil

- Issues and challenges facing urban garden continuity
- Products and functions of the garden (food, education, awareness, sociability, ecosystem services including pollination, happiness!)
- Contribution of UPA to the city's green infrastructure
- Evolution of public policies, and
- Perceptions of the future of UPA in their city

8.6 Theoretical Framework

Social regulation theory views organizations from an alternative standpoint and follows the postmodern academic tradition of deconstruction (De Terssac, 2012) while questioning determinisms, including the imposition of centralized power. By studying local arrangements of power, the autonomy of actors and how they cope with existing regulations, social regulation theory allows for a better understanding of the unpredictability, opportunism, and instability that exist at the heart of organizational systems. New relations and opportunities emerge from the interactions between actors at the heart of organizations: The actions of a group and new rules begin to take form during their meetings, conversations, and the time spent together (Idem). We need to distinguish the rule that is displayed (rule of law) and the *de facto* rule (negotiated) (Idem). This does not mean that new behaviors replace legal rules, but the latitude they have will enable certain actions, or not. "Individuals do not dissolve into a group and they act because the actions make sense to them[4]", according to De Terssac (2012: 8).

In organizations such as urban gardens, actors interact and recreate a social environment that has its own set of rules, even if they are applied at a local level only. A pattern of effective rules takes form and guides the group's life and actions (Gravel and Lemieux, 2022). It could also be used to resolve conflicts or mediate an issue. External interactions, or even external actors, are also managed to a certain extent since the group will maintain its guidelines, which are closely related to the values and principles it defends, regardless of its level of organization. Sometimes, *manifestos* are the foundation of these guidelines, as is the case with the *Horta das Corujas* (Owls Garden) in São Paulo, which we will study in more detail thereafter. Manifestos clearly express the position of their members and what they ask from the different levels of government concerning the capacity to eat healthy, organic fresh food from short food supply chains, so they win back their power to eat according to their preferences (Cockrall-King, 2016; Kalfagianni & Skordili, 2018). A clear trend focuses on the importance of gaining autonomy vis-à-vis food industrial systems often dependent on industrial complexes producing chemicals that are considered harmful to the health of the communities, the local environment, and consumers (Horta das Corujas' manifesto (2012) in Nagib, 2016).

Social regulation theory studies the existence of rules and actors' autonomy within the normative system. The rules emerging from negotiations between the actors and the system are political in nature. The most common example of this is a collective agreement, which is the result of negotiations between two parties (the employees and the organization they work for) who agree to write the rules governing the different functions of the organization. The parties together define the rules that will be used to establish the salary scale, but they will not negotiate their

[4] Translated from French by the author.

salaries. This set of "rules emerging from the negotiations, their elaboration and renewal" is what we call "regulations," and it is the subject of this theory (De Terssac, 2012: 8).

In the context of UPA in Brazil, it will be possible to differentiate the rules that are displayed (rules of law) and the *de facto* rules (negotiated) and the impacts of this dichotomy on urban gardeners. While there is a framework of rules managing UPA initiatives in the large cities of Brazil, its implementation is slow and does not seem to follow the country's pace of development. The way municipalities reply to requests for the formalization of urban gardens is often inadequate, as revealed by several urban agriculture leaders who were interviewed. Some of them had to wait 8 years to receive legal authorization! These processing delays encourage actors to take action autonomously and to sometimes "illegally" occupy urban lots, often public vacant lots, to carry out their activities.

Nevertheless, the autonomy of certain actors holding key positions in organizations can create unexpected opportunities and make some pending requests progress. SRT theory studies the leeway that some actors obtain by making personal or group decisions to adjust when confronted with obsolete or incoherent rules in each situation or when they want to fill a gap or a need (De Terssac, 2012). That illustrates how the SRT reintroduces actors as system-building agents: Despite the values and political parties holding the power, actors have the freedom to choose (Idem). Le Velly and Bréchet (2011) applied this theory to better understand the organic food supply in school cafeterias in France and showed the way to a creative renewal of organizations. This consideration will be useful to understanding the elaboration process of UPA policies and the interactions between the actors who participate in the decision-making process.

8.7 UPA: Addressing Hunger in Brazilian Communities

Brazilian cities typically adhere to a territorial organization model inherited from the Portuguese colonial era. This model fosters inherent inequalities, where advantages such as centrality, accessibility, and proximity to the seashore are reserved for the upper class. Conversely, underprivileged groups often reside in valleys, flood-prone, peripheral, or steep areas. Consequently, they face increased exposure to natural hazards and climate changes, such as landslides, flooding, and stagnant waters with disease-transmitting insects. These areas also contend with contaminated air, accumulated waste, and diseases. Moreover, they are often peripheral to urban core areas that boast better public services and lower vulnerability levels. These vulnerable urban communities often experience what Dubbeling et al. (2019: 33) call "silent hunger," which is a result of micronutrient deficiencies and food insecurity. City administrations worsened living conditions by prioritizing a globalized economy and industrial production over social and environmental concerns such as social inequalities, conservation of vegetation, water protection, and improving living spaces (Mougeot, 2005b: 110; Rivière d'Arc, 2006). Geographic mobility issues in metropolitan areas also affect many urban citizens' ability to secure and maintain employment, access basic services, and obtain food. The same holds for the transportation and distribution of food to supply urban food retailers.

For instance, exploring São Paulo's municipal market, located adjacent to a major highway and overseeing the daily redistribution of 350 tons of food, employing 1500 individuals (Municipal Market of São Paulo, website), offers valuable insights into the food supply challenges faced by this metropolis. The smooth flow of food products through distribution chains connecting agricultural regions with large cities presents a recurring challenge, requiring meticulous organizational and transportation logistics (FAO, 2007: 1).

During crises, pressures on production and distribution systems escalate, potentially halting food supply to marginalized populations. Climate change intensifies droughts, disrupting agricultural cycles, while flooding can impede food distribution networks. Hence, implementing strategies to enhance urban food resilience is crucial. Leveraging vacant lands like valleys for fresh fruit and vegetable production is essential, particularly when these spaces can be safely utilized, ensuring equitable access for people of all ages and genders. Prioritizing urban farmers' safety is crucial in urban garden planning, especially in areas affected by crime. Associative forms of urban gardening enable workers to establish a watch system and implement garden fencing for added security.

Since the 1990s, Brazil has initiated food security programs at the municipal level to address malnutrition among urban populations. These programs have also provided advantages to local food producers seeking consumers for their products (Bricas & Conaré, 2019: 9). Partnerships have been forged to supply school cafeterias, where it became compulsory for schools to purchase a designated quota of local and organic products, as well as to support soup kitchens and food banks, as elaborated further (Idem). Following his first accession to power, Brazil's president, Luiz Inácio "Lula" da Silva (2003–2011; 2023-present), took inspiration from these projects to create Brazil's Zero Hunger program, which aimed at eliminating hunger and promoting urban agriculture and family farming (Ibid.). That program was created to mitigate the consequences of poverty and give people in need some tools to build a decent life for themselves (Hall, 2006).

While the legal framework supporting urban and peri-urban agriculture (UPA) in Brazil is well established, the institutional structure necessary for national implementation has not progressed at the same pace. Municipalities often possess guidelines conducive to successful UPA implementation but lack the resources—financial, technical, human, or political—to actualize their projects.

8.8 The History of National and Municipal UPA Policymaking in Brazil

First, it is important to highlight the fact that even though urban and peri-urban farmers have been part of the history of Brazil since the colonial era, the first municipal initiative to encourage UPA began to emerge only at the end of the 1980s (Arlego Paraguassu, 2013: 89): The City of São Paulo innovated by integrating products from urban gardens in the urban economy with the implementation of a program at the end of the 1980s. At the national level, UPA was first supported at the beginning of the twenty-first century (2003) through the Brazilian policy for poverty reduction (Castelo Branco & de Alcantara, 2011).

Despite the existence of rules regulating UPA in Brazil, many urban gardeners still work outside of this framework because rules are not yet officialized or entirely applied by municipalities. According to Castelo Branco and de Alcantara (2011), only a few municipalities had adopted UPA regulations in 2010 (6 out of 80 municipalities surveyed), with the state of Minas Gerais pioneering proper UPA legislation (Santandreu & Lovo, 2007).

Brazil's UPA strategy was institutionalized in 2005 through the National Food and Nutritional Security Plan (Zero Hunger program) (Ibid; Arlego Paraguassu, 2013: 105). The Ministry of Social Development and Fight against Hunger (MDS) coordinated this plan during the administration of President "Lula" of the Workers' Party. While it focused primarily on fighting poverty and hunger in rural areas, it was funded by the National Program to Strengthen Family Farming (PRONAF), which is altogether present in urban, peri-urban, and rural fruit and vegetable production (Castelo Branco & de Alcantara, 2011; Nunez Sanchez et al., 2021).

In the wake of these initiatives, the government created the Programa de Adquisición de Alimentos (PAA) (Food Acquisition Program) in 2003 followed by the National Urban and Peri Urban Program in 2004. The PAA aimed to consistently provide school-aged children with locally grown quality (often organic) food. It had a second objective of creating markets for products that come from small family farming. Organic farmers were offered incentives to participate in the program (Bateman & Brochardt, 2013). While PAA has been a success in all Brazilian states,[5] benefiting over 270,000 families (as of 2012) (idem: 22), it has lost strength since 2016 with the political upheaval that the country underwent (with Dilma Rousseff's impeachment) and ensuing budgetary cuts in tune with the neoliberal globalization politics that then became predominant (Shappo, 2021: 38).

The National Urban and Peri Urban Program (2004) primarily focused on metropolises facilitating the organization of public urban gardens by funding

[5] The PAA stimulated the economy and restored pride to small farmers through frequent and guaranteed sales. It encouraged the creation of cooperatives, promoted self-management, and stressed the importance of the involvement of women in these organizations and their management while providing school-aged children with diversified meals that were prepared in schools with fresh, local products (Bateman & Brochardt, 2013: 6).

purchases of agricultural inputs and equipment and training urban gardeners. Again, it is in the state of Minas Gerais that the program was the most successful (Santandreu & Lovo, 2007). In 2006, there were already over 600 UPA projects in the country[6] (Bateman & Brochardt, 2013: 4). Since 2010, 23 metropolitan regions have benefited from this program (Ibid). Today, UPA initiatives can be found in all Brazilian states, but they are more present in the states of Minas Gerais, Rio de Janeiro, Parana, Goiás, Pará, Piauí and Bahia. UPA products—geared toward local consumption—and their structure vary from one region to another and sometimes include partnerships with local NGOs and the private sector, depending on their own local arrangements (Ibid).

It must be said that the municipality of São Paulo, the first city to adopt specific legislation in support of urban and peri-urban agriculture in 2004, with its law giving rise to the Urban and Peri Urban Agriculture Program (PRO-AURP), had among its objectives the fight against hunger, food production for self-consumption, and the promotion of ecological transition (de Lima Caldas & Jayo, 2019; Giacchè, 2016; Giacchè & Porto, 2015). This initiative of the then mayor of São Paulo, Fernando Haddad (2001–2004), anchored urban agriculture policies in the Ministry of Labor and Entrepreneurship and allowed for the creation of the first shared urban gardens. It should be noted that the rest of the MRSP has not shown equal desire to embrace the agroecological transition, meaning that the other 38 municipalities of the agglomeration have adopted UPA at a variable pace. The RMSP is considered the largest city of the Global South, with almost 21 million inhabitants, and the most crowded city of the Western Hemisphere (Prizendt, 2017).

In 2010, the São Paulo PRO-AURP was transferred to the Ministry of Supply and Environment. The Guarapinga seal was created to guarantee the quality of UPA products; it facilitates the marketing of products and ensures consumer safety (Ibid). Certified production comes from gardens that are monitored and often located on Environmental Protection Areas (Portuguese: Areas de Proteçao Ambiental, APAs) located within the São Paulo Green Belt. Organic production is mandatory in these areas (interview, director of São Paulo Green Belt Biosphere Reserve, March 2016) given the presence of water sources that are used to supply water to the metropolitan region (Braga Moraes et al., 2018, Rodrigues et al. 2020) (Fig. 8.2).

In the state of Minas Gerais, its capital city Belo Horizonte adopted a law in 2011 that established a UPA municipal policy bearing the following objectives: to ensure the population's food and nutritional security,[7] to create employment, and to generate income (Idem: 106). In the city of Salvador (state of Bahia), the 2008 Urban

[6]In 2008, there were 14 UPA support centers (CAAUP) managed by the Ministry of Social Development and Fight against Hunger (MDS) and responsible for supporting UPA initiatives in the country's metropolitan regions (Bateman & Brochardt, 2013: 4; Arlego Paraguassu, 2013: 90). Support centers offer training programs and technical support to encourage the development of UPA as a key element to ensure food security.

[7]Delisle defines food security as "the concrete translation of the fundamental human right to food and a key element of the quality of life" while nutritional security adds to this definition the access to drinking water and basic health care.) (Delisle, 1998: 309).

Fig. 8.2 A view of the metropolitan region of São Paulo and its Green Belt from the Pedra Grande lookout in Cantareira State Park. (Credit: Gravel, 2015)
Description: If it were not for the visionary ability of its founder Albert Löfgren, the Swedish naturalist and botanist, the Cantareira State Park, founded in 1868 to protect the water source in the Serra da Cantareira supplying the city, would probably be urbanized as the rest of the city. The Instituto Florestal de São Paulo (São Paulo Forest Institute) was built in the adjacent Albert Löfgren State Park, commonly known as the São Paulo Forest Gardens (*Horto Florestal de São Paulo*), north of the metropolitan region, where my visiting stays took place

Development Master Plan (law 7400/2008) considers "UA as an axis of the National Food and Nutritional Security Plan and a guideline for environmental education." (Municipality of Salvador, 2008 in Arlego Paraguassu, 2013: 106–107) The city of Salvador's organic agriculture law was approved in 2008 to support agroecological gardening on public and private lands.

Many of the urban gardens in Brazilian metropolises are located in areas that have been vacant or unoccupied for over 10 or 20 years and were the subject of clandestine appropriation first by groups of citizen activists before being legally recognized. Castelo Branco and de Alcantara (2011) found that the residents who took part in UPA activities tended to be less wealthy and educated, especially those living at the periphery where slums are numerous. Nevertheless, this is not always the case: The opposite can be observed in inner city locations, such as the case of the Horta das Corujas, which is in a well-to-do neighborhood of the municipality of São Paulo and has members who are professionals with a higher level of education (interview with a co-founder of the Horta das Corujas, March 2016). Those gardens, smaller in size, are more oriented toward agroecology education, lifestyle, and deeper social transformation and are considered part of the "visibility" category instead of the "scale garden" category (de Lima Caldas & Jayo, 2019). Whatever the socioeconomic context, urban community gardens are seen as contributing factors to the improvement of the quality of life of their local community, especially through the occupation of vacant lots and the reduction of the incidence of diseases such as the Zika virus (by eliminating stagnant waters where mosquitoes and other vector insects reproduce), the improvement of food security and urban landscapes, and, finally, the strengthening of the social fabric.

In 2012, the National Policy of Agroecology and Organic Production (PNAPO in Portuguese) and the Plan for Agroecology and Organic Production (PLANAPO) were launched (with a UA component), making Brazil a "pioneering nation in the implementation of public policies focused on agroecology" (Nunez Sanchez et al., 2021, n.d.). The following year, the State of São Paulo launched the Organic São Paulo Plan that supports organic fairs in parks and the production of compost in the city (Arlego Paraguassu, 2013; Nagib, 2016).

A bill aimed at creating a legal framework for UPA at the national level was proposed in 2015 and approved by the Chamber of Deputies in 2017 (Government of Brazil, website, 2015, 2017). As of 2023, it has yet to be approved by the Senate. In bill 906/15 presented by the Federal Deputy Father João, from the *Partido dos Trabalhadores* (Workers' Party) (PT) of Minas Gerais, UA is defined as "agricultural and livestock production taking place within a city and integrated to the ecological and economical urban systems, intended for food production, as well as the creation of other products for own consumption or to be marketed at local scale.[8]" (Idem).

The objectives of the proposed bill are the following (Idem):

- To broaden food and nutritional security to vulnerable urban families
- To encourage the occupation of urban vacant lands
- To generate alternative sources of income and occupation for urban populations, and
- To connect urban food production to institutional food acquisition programs in schools, prisons, hospitals, nurseries, community kitchens, etc.

Nunez Sanchez et al. (2021) studied the impact of the Brazilian political transition on organic producers in the country and found that registration rates of new organic production fell from an average of 43% between 2015 and 2017 to 4% after 2020 because of shifting priorities and politics toward agriculture (Idem). The following governments, run by interim president Michel Temer and then by Jair Bolsonaro (elected in 2018) from the Liberal Party, favored the pesticide industry and export-oriented agrobusiness following neoliberal premises (Idem; Shappo, 2021). These preferences came into contradiction with the agroecology movement

[8] Translated from Portuguese by the author.

that had been encouraged by the state until 2016, compromising the place Brazil occupied as a world leader in family agriculture and food security (Idem). Even during the COVID-19 pandemic, the focus on exports continued, threatening national food supplies and access to food with consequential high prices for "rice, beans, and soybean oil, that affected the Brazilian population and aggravated the problems of hunger and food insecurity" (Shappo, 2021 cited in Nunez Sanchez et al., 2021, n.d.).

Decentralization of power and close cooperation between the different levels of government are necessary to implement a national UPA policy, meet local needs, and receive financial and technical support from the ministry that will be responsible for its management. According to Arlego Paraguassu (2013: 107–8):

> The main obstacle to the implementation of an effective UA public policy is its integration into urban planning schemes, given that most Brazilian cities do not have a structured land use strategy while they are subject to pressures from the real estate market, which creates an increase in the value of urban lands and makes it difficult to use them to carry out ecological regulation activities of urban ecosystems.[9]

Before reviewing the key elements in the elaboration of UPA public policies, we will view the following urban gardens in more detail: the structure of their organization, the challenges they face, and their achievements.

Examining various case studies will shed light on the disparity between idealized norms (laws) and actual norms (practice), drawing from Reynaud's Social Regulation Theory (1997, 1999). Additionally, it will explore how actors endeavor to influence decision-making processes at the local level to realize their UPA projects.

8.9 The Metropolitan Region of São Paulo (MRSP): A Gardening History

Between 2014 and 2016, visits were conducted to meet various coordinators and gardeners engaged in activities across ten distinct urban gardens in the RMSP. These gardens exhibited diverse characteristics: some cultivated crops directly in urban soil or raised beds, while others were situated on rooftops, such as those atop a shopping mall with a green roof or created by a financial corporation atop a high-rise building. Gardens varied in ownership, ranging from private to public and community-managed, with some operated by NGOs and focusing on family agriculture for income generation. Notably, this study exclusively considered gardens located in public urban and peri-urban areas, totaling seven sites (Table 8.1). The following study sites were included in the data sample.

Despite the passage of a 2006 law in São Paulo aimed at providing a legal framework for Urban Agriculture Policies (AUP), the movement toward organic agriculture in urban areas didn't truly begin until 2010. This shift occurred when the first

[9] Translated from Portuguese by the authors.

Table 8.1 Study sites for urban and peri-urban agriculture in Brazil

Study sites for urban and peri-urban agriculture in Brazil			
Metropolis	Study sites	Activities	Accompaniment
Metropolitan Region of São Paulo and City of São Paulo	Horta[a] das Corujas, Pinheiros; Horta Comunitária da Saúde (Saúde district); Horta do Centro Cultural São Paul (Central district); Horta do Parque Linear da Integração Zilda Arns (Sapopemba district); 2 sites of Cidade Sem Fome located under power lines (Eastern Zone, São Mateus district) Collective garden under power lines of São Bernardo do Campo	Observation on production sites, interviews with UPA leaders and activists, attendance at garden activities, visit of an organic fair located in an urban park	Co-founders and founders of all cited gardens; São Paulo Green Belt Biosphere Reserve director; NGO Cidade sem Fome [City without hunger] founder and coordinator; NGO S.O.S. Abelhas sem Ferrão [S.O.S. Stingless Bees] founder; Religious association and the Brazilian Green Party, and 6 other UPA coordinators
Salvador de Bahia	Urban community gardens in the Vermelho river valley with agroforestry; Private urban family gardens with income	Urban community gardens in the Vermelho river	Leader of UPA at the Ministry of Health, city of Salvador and activist
Florianópolis	Peri-urban collective garden (in the making); inner city abandoned shared garden	Visit of production sites; visit of urban, organic fair at the Federal University of Santa Catarina, discussion with graduate students, participation to UPA policy launching meetings	SEMEAR Floripa Network, a professor and his graduate students from the Federal University of Santa Catarina, community organizers, local community organizers

[a]Horta means garden in Portuguese.

composters were installed at Horta das Corujas, as noted in an interview with Claudia Visoni, co-founder of Horta das Corujas, in 2016. Subsequently, in 2012, the city hosted its first workshop on organic waste management, and by 2015, Mayor Fernando Haddad oversaw the construction of São Paulo's first organic waste treatment facility. The facility aimed to collect and process organic waste from food fairs in the city, with the compost produced intended for public gardens. Additional facilities were constructed in 2016. Access to organic matter, particularly compost, is crucial for successful Urban Agriculture Policies (UPA). In shaping the organic waste recovery policy, the two co-founders of Horta das Corujas wielded influence within local chamber of commerce and neighborhood associations. Consequently, their garden directly benefited from organic compost deliveries.

The Metropolitan Region of São Paulo (MRSP) boasts a large network of urban gardeners who drive the movement known as *Hortelões Urbanos* (Urban Gardeners), which has been active on social media since its inception in 2011 by Claudia Visoni, a journalist and politician, along with her colleague. They are also members of the Association of Urban Gardens of São Paulo (Uniáo de Hortas Comunitárias de São Paulo), co-founded by the late Sergio Shigeeda, founder of Horta Comunitária da Saúde, to whom this chapter is dedicated. In 2024, the Hortelões Urbanos group boasts 86,9 K members on Facebook, drawing participants from Brazil and beyond, fostering connections among individuals and associations sharing an interest in Urban Agriculture (UA) and agroecology. Serving as a public forum, the network facilitates the exchange of knowledge and best practices for promoting healthy eating and cultivating green spaces within cities (Nagib, 2016: 224). Since its inception in 2012, the group has monitored the activities of two gardens situated in public areas and established the Horta do Ciclista (Biker's Garden) in the heart of São Paulo, along the bustling Avenida Paulista (Nagib, 2016: 333; Giacchè, 2016). While not included in this research due to its relatively small size and lower activity during our visit, the group, while potentially toning down its initial confrontational stance, remains committed to advocating for citizen engagement in reclaiming urban spaces and promoting pesticide-free, healthy eating.

8.9.1 Horta das Corujas, São Paulo

In the case of the Horta das Corujas, creating a welcoming space where everyone can gather, young and old, and learn about agroecology became the main motivation to transform a vacant, abandoned piece of land in the posh neighborhood of Vila Beatriz (see Fig. 8.3) where gangs used to gather at night. With the help of the Hortelões Urbanos Network volunteers, that site has been completely redesigned to provide its visitors with tranquility and safety. The green space surrounding the community garden (itself surrounded by a low fence) is often used for family gatherings, reading or yoga sessions, and school group activities, taking place under the shade of mature trees (Fig. 8.4). A small stream, *Córrego das Corujas* (The Owls stream), flows through the park and gives its name to the Horta das Corujas. It offers a rare opportunity to see a surface stream in this town since all the others have been buried to allow for urbanization.

Visitors at the garden are guided by signs explaining the dos and don'ts and the whereabouts. Participants are free to harvest whatever they feel tasty but are clearly warned not to uproot the whole plants (but some cannot resist the temptation apparently!). Its governance relies on the cofounders' vision and calls for voluntary work on an ad hoc basis. Some partnering associations are welcomed to showcase their work in situ, such as *S.O.S. Abelhas sem Ferrão* (S.O.S. Stingless Bees), founded by Gerson Pinheiro (see Fig. 8.5), and whose mandate is to rescue threatened wild native bees and educate the population about pollination services. Some wild bees lodging inside tree trunks are threatened when the municipality decides to cut trees,

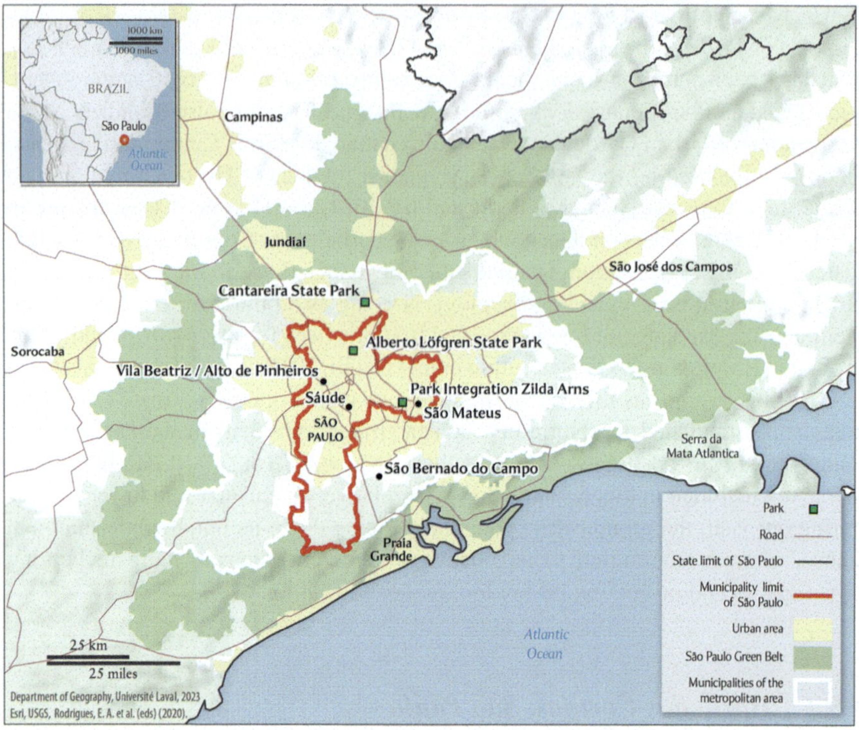

Fig. 8.3 Map of the Green Infrastructure of the Metropolitan Region of São Paulo

and this is how the NGO was born. Its founder carries the hives to the community gardens in the city to improve pollination and give the bees a new home (Interview with Gerson Pinheiro, 2015, 2016). Along with the São Paulo-based MUDA (Urban Agroecology Movement) and the Hortelões Urbanos movement, it is contributing to the ecological transition of the RMSP.

8.9.2 Cidade Sem Fome Community Gardens

Primarily utilizing land under power lines in the eastern periphery of São Paulo (Zona Leste), the NGO Cidade sem Fome (City without Hunger) has provided vacant lots for families seeking income through agriculture. Since its establishment in 2004, the NGO has established 25 community gardens and 38 school gardens in this area of São Paulo, renowned for its low-income neighborhoods and housing approximately 3.3 million people, with 40% unemployed (Cidade sem Fome website; Giacchè, 2016). An interview with founder Hans Dieter Temp, along with on-site visits accompanied by an NGO representative, allowed for an assessment of the team's work and the enthusiasm of around 150 urban farmers benefiting from this

Fig. 8.4 A view from Horta das Corujas. (Credit: Gravel, 2016)

initiative. Many of these farmers are retired couples and families who supplement their income through direct sales of fresh produce and eggs (Figs. 8.6 and 8.7).

8.9.3 São Bernardo do Campo

In the southeastern part of the RMSP, within the ABC Zone of Greater São Paulo, one can find collective gardens overseen by a religious NGO and representatives of the Brazilian Green Party. During our visit guided by the two coordinators, we observed the remarkable success and expansion of the project since its inception in 1984. Initially a personal endeavor by a local psychologist during his leisure time, the project has since grown to encompass 12 gardens spanning 2.8 hectares under power lines. The religious NGO has assumed responsibility to meet the needs of this historically industrial neighborhood. Interviews revealed that in the past, urban and peri-urban agriculture (UPA) in São Paulo was not associated with leftist politics, but rather with conservative and religious parties aiming to assist needy residents.

These 12 gardens now cater to the needs of 70 gardening families, who supply their neighborhood with approximately one ton of organic produce annually, including fruits, herbs, vegetables, and eggs, sold directly by the roadside (Pereira, 2019).

Fig. 8.5 A view of a rescued wild bee colony managed by S.O.S. Abelhas sem Ferrão and its director, Gerson Pinheiro (left) in the Horta das Corujas. (Credit: Gravel, 2016)

The gardeners' diverse age groups present during our visit underscored the garden's contribution to the well-being of senior citizens. One coordinator explained that many elderly individuals feel isolated in their apartments and long for the rural lifestyle they once knew. The municipality's name, "do campo" (meaning "of the countryside"), attests to its rural identity, with the area once occupied by farms (*fazendas*). Regular visits to the urban garden allow seniors to overcome their isolation and find renewed purpose in life. Moreover, garden coordinators have adapted infrastructures to accommodate their physical capabilities.

The São Bernardo gardens were recognized by the municipality in 2016 for their contributions to sustainable development and efficient water use. Unlike collective gardens where one large plot is cultivated by all gardeners, the structure of community gardens allows each family to cultivate their plot according to their needs and preferences. Here's an example of an agroecological garden plot, nicknamed "todo misturado" (all mixed up) by fellow gardeners, who sometimes lament the lack of straight lines and order of the plants (Fig. 8.8). This agroecological garden style, originating from the state of Bahia, emphasizes the natural complementarity between plants, a fundamental principle of agroecology. Notably, both pictures from the same site feature fruit trees, including avocado (as seen in Fig. 8.9). This approach promotes enhanced production and plant health while minimizing soil depletion compared to the well-ordered Western-style garden (Fig. 8.10).

Fig. 8.6 Cidade sem Fome garden under the power lines, São Mateus, eastern zone of the Greater São Paulo

8.9.4 Horta Comunitária da Saúde (Health Garden)

The small Horta Comunitária da Saúde began in 2013, and its leader waited for 8 years before obtaining municipal authorization in 2021.[10] During this time, he and his fellow gardeners remained active, organizing various activities in the 0.04-hectare garden located on a dead-end street and installing rainwater collection and composting equipment (Forbes Brazil, 2022). Situated in the inner city of São Paulo, in the Saúde district (Health district), a mixed-income neighborhood south of the RMSP, this community garden reflects the dedication of its late leader, Sergio Shigeeda. Starting with clearing a space filled with garbage on his street, Sergio and his fellow gardeners gathered to assist him[11] (Interview with Sergio Shigeeda, 2015). Despite working in computer sciences professionally, Sergio was a self-taught gardener, activist, politician, and spiritual individual. He shared his expertise with numerous

[10] This decision coincided with the City of the São Paulo' objectives for 2021–2024 to create 400 inner city gardens (City of São Paulo, 2021: 48).

[11] To see an aerial view of the Saúde Garden, see Forbes agro (2022) https://forbes.com.br/forbesagro/2022/01/no-aniversario-de-sao-paulo-suas-hortas-urbanas-mostram-o-campo-na-cidade/

Fig. 8.7 Cidade sem Fome gardener showcasing her garden and trees under the power line, São Mateus, eastern zone of the Greater São Paulo. (Credit: Gravel, 2016)
Description: In Fig. 8.7, it is possible to see the wide array of vegetables and fruit trees cultivated by this woman and her aged husband (not shown). Papaya, banana, and coffee trees can be seen in this picture

followers, teaching them about plant knowledge, unconventional food plants,[12] permaculture,[13] organic soil amending practices, and rainwater collection with cisterns. His primary motivation was to mitigate the impact of human activities on the environment, and he was passionate about safeguarding all forms of life. I had the privilege of meeting Sergio Shigeeda in 2015, and we continued our exchanges for 7 years after my return to Canada. Sergio introduced me to the Moringa tree—also known as the "Tree of Life" or "Miracle Tree"—emphasizing its various benefits for humans and animals[14] (Sutton, 2021). Originating in India, Sergio believed that this tree could potentially save millions of people. He pointed out one such tree at the edge of his garden, near the street, when bidding farewell.

[12] The term unconventional food plants apply to plants or part of plants (such as carrot foliage, for instance), that are not usually consumed by humans. Many visibility gardens in the MRSP showcase these plants as part of their educational program.

[13] Sergio Shigeeda discovered permaculture and applied its principles to manage the residuals of gardening operations without energy losses.

[14] To read more on the Moringa tree, see Samantha Sutton's article in reNature magazine.

Fig. 8.8 Example of a Bahian agroecological style plot in the community garden of São Bernardo do Campo. (Credit: Gravel, 2016)

Sergio Shigeeda initiated numerous tree plantation activities, known as *Plantios Globais* (Global Plantations), which gained traction through social media and attracted hundreds of participants in São Paulo, other Brazilian cities, and around the world. This initiative culminated in 100 cities worldwide joining the Global Tree Plantation Program. In 2017, in support of pollinators, 1000 São Paulo residents gathered to plant the Green Melliferous Corridor (*Corredor Verde Polinizador*) with native melliferous trees on the property of the Biological Institute, following Sergio's lead. Partnerships for this endeavor included the Municipality Green Secretary, the State of São Paulo Ministry of Agriculture and Supply, the Biological Institute, the Subprefecture of Vila Mariana, and local NGOs (Shigeeda's website; follow-up interview in 2018).

During the COVID-19 pandemic, Sergio was actively involved in distributing organic produce baskets to needy families, particularly in the eastern part of the RMSP. He also advocated for the continuation and revitalization of organic fairs in São Paulo, where community garden produce was sold (City of São Paulo, 2022, website). Throughout these endeavors, Sergio battled cancer, and his tree plantations and community garden stand as a testament to his love for the city of São Paulo. He passed away in 2022 and was posthumously honored by the Municipality of São Paulo for his environmental actions.

Fig. 8.9 Cultivation with fruit trees under electric lines, São Bernardo do Campo. (Credit: Gravel, 2016)

There are several other noteworthy urban and peri-urban agriculture initiatives in the RMSP, such as the Horta Zilda Arns, established in 2010. Located in the Zilda Arns Integration Linear Park in the eastern part of the RMSP, this park is the third longest linear park in the world. Although I have met with the founder of the Zilda Arns Garden, André Manoel Dos Santos, on several occasions, our interactions have always taken place outside the garden. Therefore, I am not including it in my analysis here due to a lack of firsthand information and observation.

8.10 Salvador de Bahia

One of the most innovative forms of UPA in Brazil that I witnessed can be found in the valley of the Vermelho River, flowing through the city of Salvador. On the inclined slopes owned by the state's electricity company, citizens cultivate the land with the company's permission. The municipality divided this rugged terrain into several parcels and allocated them to interested families. Each parcel is managed by a family who grows vegetables under fruit trees in tropical agroforestry systems. This practice is widespread in Bahia, contributing to the state's reputation within the Brazilian gardeners' network as a major supplier of organic produce and seeds. The

Fig. 8.10 Western-style gardens in São Bernardo do Campo. (Credit: Gravel, 2016)

crops are sold at roadside stands (see Fig. 8.11), providing neighboring residents with easy access to "good, clean, and fair food" (the Slow Food motto) (Petrini, 2007) at affordable prices.

The fruit trees (e.g., avocado, bananas, nut trees) growing on the slopes help protect the riverbanks, provide shade for leafy greens, control erosion, and retain sediments before they reach the river. After studying the potential health risks associated with UPA, especially concerning car exhaust pollution on leafy vegetables, the city's Ministry of Health decided to support this practice through public policy (Interview with Lídice Arlego Paraguassu, 2016). Consequently, a regulatory framework was established to monitor gardening activities throughout the city, including vacant urban and private lots.

The UPA leader in Salvador de Bahia, Lídice Arlego Paraguassu, combines two roles: she is a university biology professor and a public servant at the City's Ministry of Health. Her doctoral thesis on UPA, completed in 2013, explored its potential to increase sustainability in Salvador. Her research demonstrated a link between the population's well-being (food security, quality of life) and environmental protection (conservation of valley bottoms and steep lands) through UPA public policies.

Several factors contributed to the success of her proposal. The Ministry of Health hired researchers to study urban soil quality and food product safety. Local companies, including the power company, provided support. An accreditation system was established to allow the sale of UPA products at roadside stands, which promised to

Fig. 8.11 Fresh fruit kiosk adjacent to a garden in Salvador de Bahia. (Credit: Gravel, 2016)

diversify the diets of low-income urban populations. Farming families are selected by the Ministry of Health, and UPA is recognized as an economic sector that contributes to household incomes in Salvador de Bahia.

8.11 Florianópolis, Santa Catarina

The city of Florianopolis, in the state of Santa Catarina in southern Brazil, stood out with promising initiatives in urban agriculture in 2016. The first partners behind this movement were students and professors from the CEPAGRO (Center for the Study and Promotion of Group Agriculture) at the Federal University of Santa Catarina. They formed the SEMEAR Floripa Network (SEMEAR means "to sow" in Portuguese) in collaboration with the city's Ministries of Health and Education. Since 2017, Florianópolis citizens have enjoyed increased access to fresh and organic vegetables and herbs from more than 100 gardens cultivated on public lands, thanks to the Municipal Urban Agriculture Program (PMAU). Situated on an island near the Brazilian Atlantic coast, Florianópolis must adhere to a state law that mandates organic agriculture; the use of chemical inputs is prohibited on the seacoast of Santa Catarina to protect marine fauna and flora.

Weekly meetings allowed SEMEAR's partners to develop a shared vision of agroecological food systems where urban and peri-urban agriculture thrive (Rede SEMEAR Floripa website). Their collective efforts aimed at defining a public UPA policy, building on the lessons learned with the PMAU (City of Florianópolis, 2019, website). My assistant, Josée Lemieux, and I visited some of these gardens on the outskirts of the city (Fig. 8.12) and in the inner city, as well as accompanied the members of the SEMEAR Floripa network to the first meeting to launch the joint initiative. We also attended one of the organic fairs organized every Sunday on the University Federal of Santa Catarina (UFSC) campus (Fig. 8.13). Alongside gardeners, there were beekeepers, cheese makers, and bakers offering products made in harmony with nature. The university has played a key role in mobilizing UPA partners through CEPAGRO, which has existed for over 30 years. This center brings together students and researchers from agricultural sciences, education, and other fields interested in agroecology and urban agriculture. Some graduate students

Fig. 8.12 Raised beds at the periphery of Florianopolis ready for plantation. (Credit: Gravel, 2016)

Fig. 8.13 The author and a student vendor at the Sunday Organic Food Fair at the Federal University of Santa Catarina. (Credit: Gravel, 2016)

became activist gardeners and got involved in municipal politics, which strengthened the movement.

During my visit in 2016, the SEMEAR Floripa movement was in its infancy, having been created in 2015. Follow-up interviews revealed that 5 years later, significant progress had been made. Key advances included community gardens gaining access to local compost made from domestic organic waste, enabled by the innovative Florianópolis composting law (law n°10501/19) (Di Fiore et al., 2022), the development of public policies governing UPA, and the institutionalization of UPA discourse through a city decree (which had mixed reactions from the SEMEAR team). Additionally, the mapping of the city's UPA sites was completed (SEMEAR leader interview, October 2020).

The COVID-19 pandemic and political shifts at the national level since 2016 significantly reconfigured UPA partners and participants in Santa Catarina. The change in national leadership affected regional support for UPA after 2 years of strong development. Moreover, the pandemic exacerbated urban poverty from 2020 onward, severely impacting families dependent on daily sales. This led to an increased effort to produce more food from urban and peri-urban gardens to combat growing food insecurity among various population groups.

Follow-up interviews with SEMEAR leaders shed light on the consequences of policy changes at various levels of urban agriculture (UA) governance. As elsewhere in Brazil, the urban gardeners' demand for pesticide-free products aligns them ideologically with the political left, which is critical of agribusiness. This alignment with agroecology-supporting local politicians can benefit urban gardeners but also expose them to vulnerabilities when their political allies are in opposition. Such was the case for the SEMEAR Floripa Network, which faced marginalization due to shifts in municipal power.

SEMEAR key figures believe urban agriculture has become institutionalized, but this process has somewhat sidelined the engaged citizens who initiated and sustained the movement. A communication breakdown exists between certain representatives from the Ministries of Health and Education in Florianópolis, who control the budget and decision-making for UPA development, and SEMEAR Floripa Network members, who have advocated for UPA since 2015 and possess the necessary expertise and manpower. These actors are now divided by political allegiances, further complicating cooperation and progress in urban agriculture initiatives.

8.12 Analysis

In all of the sites selected for our research, the factors that most influenced the emergence and survival of urban gardens are:

1. Citizens' involvement at the garden's inception and their motivation to carry out the project.
2. Citizens' voluntary work in the garden's activities and operations.
3. Inclusion in an urban gardener network, whether local or virtual.
4. Local political support and collaboration for UPA.
5. An absence of land pressure due to geographical location and soil type, with the most enduring gardens situated on uninhabitable lands such as valley slopes, street dead ends, and lots under power lines.

These findings partially align with Carolino's, 2021 results, which concluded that a lack of political interest leads to difficulties and abandonment of many urban gardens in the Metropolitan Region of São Paulo. Carolino also examined how the lack of housing policy impacts the status and vulnerability of community gardens, suggesting that this deficiency explains why community gardens often occupy unused lands, garbage dumps, and areas of drug consumption, making it challenging for urban gardeners to start production. Despite these obstacles, activist gardeners are drawn to land reclamation, which attracts other gardening "guerrilleros" determined to pursue deeper social transformation by transforming and appropriating urban spaces (Visoni & Nagib, 2019).

Food production is secondary in these "visibility gardens" (de Lima Caldas & Jayo, 2019), where the focus is primarily on education, agroecology, land reclamation, and political messaging. These gardens are usually found in inner city, trendy areas populated by young and middle-aged professionals and university students, such as Horta das Corujas, Horta Comunitária da Saúde, and Horta do Centro

Cultural. In contrast, "scale gardens," (Idem) where production volumes are higher, are typically located on the periphery of large cities where more space is available, such as under electric lines in the east and southeast of the MRSP, where kilometers of gardens are managed by the NGO Cidade sem Fome, or in Salvador on valley slopes provided by the utility company for gardening. These gardens focus more on creating income sources and feeding peri-urban residents.

8.12.1 Applying Social Regulation Theory to the Case Studies

Social regulation theory, viewed through the lens of public action, recognizes the diversity of actors and their preferences, values, and choices. Urban gardeners are the focal actors in this study. They gather, cooperate, and exchange ideas in spaces where they can establish their own set of rules, such as compost management, teamwork, and social media communications. Each member contributes their experience to the group's collective reflection and can influence the decision-making process.

The survival and recognition of a UPA project are enhanced by consistent efforts, external allies (partnerships, networks, committed politicians, high-ranking civil servants), and effective leadership. UPA initiatives, regardless of political allegiance, can be justified in two ways. First, by their contribution to family income and well-being, an argument favored by members of center and right-wing parties in partnership with the Church. Second, by reclaiming urban spaces for citizens' use, promoting environmental education, and supporting autonomous organic food growing, an argument typically made by left-wing parties.

Despite the slow response and official acknowledgment from authorities regarding UPA activities, as evidenced by seven interviews conducted in São Paulo (indicative of a lack of political interest according to Carolino, 2021), these activities persist outside of official regulations until they receive recognition that grants them the "right to exist." In the case of the Horta Comunitária da Saúde, its leader sought the city's approval to utilize a vacant lot at the end of a street for 8 years before being granted authorization in 2021.[15]

While UPA activities remain largely unregulated by the state—sometimes to the benefit of the gardeners—their practice reveals modes of regulation, competing visions, and power dynamics in the negotiation of rules. For instance, the Horta das Corujas Collective protested against the widespread use of insecticides sprayed by municipal officers to kill mosquitos during the Zika epidemic in RSMP in 2015, requesting that their garden be excluded to protect wild bees and organic production. Their demands to local authorities included information on the spraying schedule and its potential impacts on human health. The city responded that the spraying routes were determined at the last minute, making it impossible to provide a 24-h

[15] This decision coincided with the City of São Paulo's objectives for 2021–2024 to create 400 inner-city gardens (City of São Paulo, 2021: 48).

warning for gardeners to protect their beehives and vegetables. Consequently, the gardeners created an alert system, where the closest neighbor would run and alert others whenever authorities came to spray. This episode was highly stressful for the gardeners and highlighted a lack of communication and understanding on the part of the authorities (Visoni, personal communication, 2023; Visoni, 2024).

8.12.2 The Social Actor and the System

It is crucial to understand the relationship between the social actor and the system. The study of discourses elucidates that actors build society, not the other way around, and that society is the product of various commitments by actors with different positions (De Terssac, 2012). Actors who rebel against the rules and choose to personally act in favor of a cause within the system aim to contest existing rules and create new ones. Consequently, they reveal a plurality of legitimate sources of regulation. Actors attempt to exercise their autonomy by gaining small "pieces" of personal leeway, as demonstrated by Le Velly and Bréchet (2011) in the case of French school cafeterias, where considerable effort was required for one actor responsible for purchasing food to introduce organic products. This dynamic is also reflected in the examples gathered through field research for this study.

Operational and cooperative rules from UPA practice ensure gardens are maintained and endure over time. Interviews with civil servants and garden coordinators revealed how they navigate and manipulate the rules, gaining leeway within the municipal system to support UPA. Individual actions included: (1) diverting funds by a municipal public servant from sewing workshops to urban gardener training; (2) leveraging influential business or civil servants to gain municipal recognition for UPA; and (3) using double institutional affiliation to protect progress and enact public policies that institutionalize UPA activities.

8.13 Discussion

8.13.1 Strategies for Ensuring the Sustainability of Urban Gardens

According to urban garden coordinators in Brazil, the creation and survival of gardens initially depend on informal acknowledgment from municipal officials, followed by formal recognition. To increase visibility, leaders often engage in neighborhood councils or municipal/state politics, securing the gardens' place in the urban landscape and garnering support. This political representation is crucial for showcasing the gardens' achievements and rallying local and influential community members. Visoni (personal communication, 2023) noted that some gardens do not survive due to community pressure to close them, driven by hygiene concerns. "For some reason, she says, some people have more faith in poison than in Nature." Urban garden leaders often face challenging times and anxieties, especially when dealing with authorities.

Urban gardeners live with uncertainty and impermanence. Efforts to achieve permanence and recognition consolidate solidarities. They mobilize individuals to green cities, initiate cleanup and tree plantation operations, and organize organic fairs. People join based on a shared desire to improve city neighborhoods and be agents of change.

8.13.2 Urban and Peri-Urban Agriculture: Political or Not?

Some view UPA activities as inherently political, linked to citizens' demands for environmental education and agroecology. Interviewed leaders felt they disrupted the established order by promoting diverse urban space uses, often being seen as ecologists, anarchists, activists, or gardening "guerrilleros." Despite existing laws, agreements between urban gardens and cities in São Paulo are mostly informal, creating vulnerability. Small volunteer groups, gardeners' age and health, high volunteer turnover, and weak loyalty were identified as issues that could weaken UPA organizations.

8.13.3 Gardens, Places of Rule Production

Regardless of their model or structure, all gardens are spaces where local rules are created through interactions. Sometimes these rules manage visitor behavior: small signs urge visitors to stay off garden beds, avoid uprooting plants, and follow compost production rules. At other times, they manage group life, even regulating sexuality among users, including seniors, who are also subject to sanctions.

Urban gardeners collaboratively establish reasonable rules to govern their shared working space, which can lead to conflicts. As De Terssac (2012: 8) explains, "interaction generates rules which are valid and legitimate at the local scale but still enable communication, social exchange, cooperation, conflicts, arbitration, and compromises".[16] The frequent presence of urban gardeners and the constant arrival of new individuals are crucial for the continuity of urban gardens, which often exist without government financial support.

[16] Translated from French by the author.

8.13.4 Citizens' Well-being and Cultural Empowerment

Here is the text made more concise:

Access to green spaces enhances citizens' physical and mental well-being by offering a break from urban traffic. UPA provides physical exercise, stress reduction, grounding through connection with nature, and overall well-being (Boukharaeva et al., 2005). As Mougeot (2005b: 276) notes, urban gardens cultivate both vegetables and human beings, reflecting a mosaic of food cultures. These gardens offer a variety of fresh produce, allowing consumers to buy niche products, such as exotic or heritage varieties.[17] Food enthusiasts and immigrants seeking ingredients for traditional recipes benefit from these offerings (Reyburn, 2002). Urban gardens, often part of cities' green infrastructure, provide accessible spaces for hobbies and sports, and offer free edible products to cyclists and pedestrians, enhancing well-being and belonging while bypassing conventional consumption channels.

8.13.5 Toward Public Policies for Urban and Peri-Urban Agriculture?

This chapter provides steps, tools, and advice to establish a participatory approach in UPA through initiatives like participatory urban planning or MPAP (multistakeholder policy formulation and action planning) as described by Dubbeling et al. (2011), which involves concerted actions among various UPA actors (urban gardeners, government levels, real estate developers, citizens' councils, etc.).

Discussions with urban garden leaders in three Brazilian Metropolitan Regions reveal a favorable framework for developing public policies that encourage UPA implementation at the municipal level. Requests and achievements by volunteer urban gardeners can help formulate core policies to further support citizens' initiatives (Mougeot, 2005b).

These achievements include connections with food acquisition programs for school cafeterias, phytosanitary regulations, support from sponsors like the power company ceding land in Salvador de Bahia, and job grants for worker reintegration and gardener training (Ministry of Labor, São Paulo). However, key issues remain, as noted by interviewees, needing official acknowledgment to:

1. Ensure continuity in land occupation
2. Promote expansion and multiplication of cultivated parcels
3. Access public funding for basic infrastructure and operating expenses
4. Obtain the right to market produce from urban gardens, and

[17] Among others, the Incredible edibles in the UK and the unconventional food plants—the PANCS, (Plantas Alimentícias Não Convencionais, in Brazil).

5. Regulate pesticide use in cities to protect crops, water quality, and pollinators

Finding urban ecosystems in the list of ecosystems assessed by the MEA can be surprising (MEA, 2005; Mougeot, 2005a), but some have rich biodiversity, such as the São Paulo Green Belt, which aims to protect the remaining 7% of the Atlantic Rainforest (Mata Atlántica) (Braga Moraes et al., 2018; Rodrigues et al., 2020). Despite urbanization, these remnants are well preserved thanks to urban parks and conservation units (Rocha & Feldmann, 2006). Urban ecosystem services could be enhanced by creating more linear parks, community gardens, revegetating riverbanks, and planting urban forests (Braga Moraes et al., 2018; Gravel et al., 2016; Dumas, 2018; Belevi & Baumgartner, 2003). Smart city greening should consider project space, height limits, and species selection, prioritizing native species for better adaptation and preference by native pollinators (Wamsler, 2015; Wamsler et al., 2013; Tallamy, 2018).

8.13.6 *The Future of Urban and Peri-Urban Agriculture in Brazil, As Envisioned by Garden Coordinators*

The future of UPA in Brazilian cities, according to urban garden coordinators, will rely on peri-urban family agriculture due to its higher productivity and larger land use compared to inner-city community gardening. Peri-urban agriculture contributes more to household income and employment. The educational and social benefits of intraurban agriculture are often overlooked by decision-makers due to its low productivity. Many politicians and citizens view UPA negatively. Community UPA in São Paulo suffers from being associated with left-wing movements, while conservative governments initially supported it in peripheral municipalities. To enhance urban food security, UPA should be depoliticized and democratized, as successfully done by the NGO Cidade sem Fome in São Paulo's periphery.

What could be the potential impacts of institutionalizing UPA? Integrating urban and peri-urban agriculture (UPA) management plans at the municipal level could end citizens' struggles and mobilizations. In Salvador de Bahia, the Ministry of Health manages UPA, while in Florianópolis, UPA development involves health sector and academic actors with a holistic vision. Once institutionalized, militant actors become urban gardeners and fresh produce consumers. This institutionalization may benefit selected citizens, reducing grassroots power unless participatory management is implemented. Brazil could use its participatory budgets as a model to create a governance framework for UPA.

UPA services and functions, such as creating several amenities that help improve the quality of urban landscapes, favor a healthy diet and lifestyle, and contribute to fighting poverty and malnutrition, should be promoted to mobilize citizens (FAO, 2014; Cockrall-King, 2016; Duchemin, 2013).

8.14 Conclusion

Although Brazil has integrated UPA into its program to combat hunger and malnutrition, many obstacles to developing and maintaining urban gardens remain. Despite underfunding and low political support for UPA projects in metropolitan areas, some actors still manage to develop initiatives and promote their socioeconomic and environmental benefits. This chapter presents attempts by UPA actors to create municipal policies to meet local needs for fresh produce and build green cities. It highlights negotiation processes that break from previous social rules and allow new norms to emerge, where market forces, local stakeholders, multilevel governance, and municipal regulations coexist. The study analyzes the progress made by UPA actors in each city and their implications for municipal public policies promoting urban agriculture.

Providing fresh, local food that meets cultural preferences enhances life quality, diet, and happiness through collective urban agriculture. This rewarding activity benefits urbanites and their families and can enhance urban sustainability. Decision-makers should preserve and create green productive spaces within urban areas. Urban agriculture, with its proximity and affordability, is a crucial tool to fight poverty and malnutrition in Brazil. Despite government efforts, urban agriculture is not yet institutionalized at the municipal level to recognize its value and functions as an essential service to urban living.

Despite informal efforts by garden coordinators, many agree that a comprehensive legal framework and urban planning are needed to democratize urban agriculture and enhance its benefits. The institutionalization of UPA in Salvador de Bahia and Florianópolis, anchored to the Health Ministry, serves as a successful model. Moving away from the iconoclast ideology of gardening guerrilleros, UPA democratization promises greater access to shared gardens for urban families and meets food needs in a depoliticized manner.

Gardening guerrilleros have highlighted the importance of citizen appropriation of urban green spaces and local food production, raising awareness about eating locally. However, the future of UPA in Brazilian cities likely lies in peri-urban areas, where families can earn a living from more stable and productive UPA activities. These areas offer space for animal rearing and fruit tree planting. Designing public policies that align with gardeners' preferences and guide the productive process is crucial for institutionalizing UPA and promoting urban sustainability and resilience.

Acknowledgment The author expresses gratitude to the Government of Canada for funding the research in Brazil through the Canada-Brazil Joint Research Project on Environmental Governance from 2014 to 2016. Special thanks are extended to the following individuals for their assistance during fieldwork and in answering questions: Pedro Jacobi, Professor at the Institute of Energy and Environment at the University of São Paulo; Louis Guay, retired Professor from the Sociology Department at Laval University; Edgar F. de Luca, former director of the Instituto Florestal de São Paulo; Malu Ghandra, former director of the São Paulo Green Belt Biosphere Reserve; Clécio Azevedo da Silva, Professor at the Department of Geography at the Federal University of Santa Catarina; Lídice Almeida Arlego Paraguassú, Professor at the Federal Institute of Bahia; and Claudia Visoni, State of São Paulo representative and leader of the Horta das Corujas. Additionally, gratitude is extended to Josée Lemieux, a former M.A. student who graduated from the Department of Geography at Laval University under the author's supervision and contributed as a research assistant.

References

Amato-Lourenco, L. F., Lopes Moreira, T. C., de Oliveira Souza, V. C., Barbosa, F., Jr., Saiki, M., Nascimento Saldiva, P. H., & Mauad, T. (2016). The influence of atmospheric particles on the elemental content of vegetables in urban gardens of São Paulo, Brazil. *Environmental Pollution, 216*, 125–134. Retrieved at: https://pubmed.ncbi.nlm.nih.gov/27249450/. Accessed 13 Jan 2024.

Arlego Paraguassu, L. A. (2013). A agricultura urbana como estratégia de sustentabilidade da cidade do Salvador, Bahia, Brasil. Universidad de Salamanca, Spain, Unpublished Ph.D. dissertation. Accessed August 30, 2024 http://hdl.handle.net/10366/122904

Aubry, C., & Consalès, J.-N. (2014). L'agriculture urbaine en question: épiphénomène ou révolution lente ? *Espaces et sociétés, 3*(158), 119–131. Retrieved at : https://www.cairn.info/revue-espaces-et-societes-2014-3-page-119.htm. Accessed 13 Jan 2024.

Aubry C., J. Ramamonjisoa, M.-H. Dabat, J. Rakotoarisoa, J. Rakotondraibe and L. R. Rabeharisoa (2012) "Urban agriculture and land use in cities: An approach with the multifunctionality and sustainability concepts in the case of Antananarivo (Madagascar)". *Land Use Policy*, vol. 29, n° 2, pp. 429-439. Retrieved at: https://www.sciencedirect.com/science/article/abs/pii/S0264837711000937?via%3Dihub. Accessed 13 Jan 2024.

Bateman, J., & Brochardt, V. (2013). *Brazil's lessons in rural development. Family agriculture, access to water, and civic engagement*. Washington Office on Latin America (WOLA). Retrieved at: https://www.wola.org/sites/default/files/downloadable/WOLA_RPT_Open-Doors_FNL_med.pdf. Accessed 13 Jan 2024.

Baudelet, L. (2015). Pollution et risque alimentaire dans les jardins partagés franciliens. *Revue d'ethnoécologie, 8*, 11 p. Retrieved at: https://journals.openedition.org/ethnoecologie/2441#quotation. Accessed 13 Jan 2024.

Bebbington, A. (2004). Livelihood transitions, place transformations: Grounding globalization and modernity. In R. N. Gwynne & C. Kay (Eds.), *Latin America transformed. Globalization and Modernity* (2nd ed., pp. 173–192). Edward Arnold.

Belevi, H., & Baumgartner, B. (2003). A systematic overview of urban agriculture in developing countries from an environmental point of view. *International Journal of Environmental Technology and Management, 3*(2), 193–211. Retrieved at: https://ideas.repec.org/a/ids/ijetma/v3y2003i2p193-211.html. Accessed 13 Jan 2024.

Boukharaeva, L. M., Kauark Chianca, G., Marloie M., Toledo Machado, A., and Torres de Toledo Machado, C., (2005) "L'agriculture urbaine comme composante du développement humain durable: Brésil, France, Russie". *Cahiers Agricultures*, vol. 14, n° 1, p. 154-158. Retrieved at: http://revues.cirad.fr/index.php/cahiers-agricultures/article/view/30506/30266. Accessed 13 Jan 2024.

Braga Moraes, R. A., Rodrigues, E. A. B., Camacho Pires, C., Luca, E. F., Gravel, N., & Chagas De Carvalho, Y. M. (2018). A escassez hídrica e seus reflexos sobre os serviços ecossistêmicos e o bem-estar humano na Reserva da Biosfera do Cinturão Verde da cidade de São Paulo. In M. Buckeridge & W. Costa Ribeiro (Eds.), *Livro branco da água. A crise hídrica na Região Metropolitana de São Paulo em 2013–2015. Origens, impactos e soluções* (pp. 54–73). University of São Paulo. Retrieved at: http://www.iea.usp.br/publicacoes/ebooks/livro-branco-da-agua. Accessed 13 Jan 2024.

Bricas, N., & Conaré, D. (2019). Perspectives historiques sur les liens entre ville et alimentation. In *Facts Report* (20) *L'agriculture urbaine, nourrir les villes autrement* (pp. 6–11). Retrieved at: https://www.institut.veolia.org/fr/publications/revue-linstitut-facts-reports/agriculture-urbaine-nourrir-villes-autrement. Accessed 13 Jan 2024.

Carolino, K. (2021). Urban Agriculture in Brazil: an analysis of public actions implemented in community gardens. *QES Climate Justice*. York University. Retrieved at: https://qesclimate-justice.info.yorku.ca/research-projects/urban-agriculture-in-canada-an-analysis-of-public-actions-implemented-in-community-gardens/. Accessed 13 Jan 2024.

Castelo Branco, M. and AF. de Alcantara (2011) "Hortas urbanas e periurbanas: o que nos diz a literatura brasileira?" *Horticultura brasileira*, vol. 29, n° 3, p. 421-428. Retrieved at: https://www.scielo.br/j/hb/a/BBhZ9hvsDdRCbwd9mQF87ZQ/. Accessed 13 Jan 2024.

Cockrall-King, J. (2016). *La révolution de l'agriculture urbaine*. Écosociété.

De Bon, H., Parrot, L. and Moustier, P. (2010) "Sustainable urban agriculture in developing countries. A review". *Agronomy for Sustainable Development,* vol. 30, n° 1, p. 21–32. Retrieved at: https://link.springer.com/article/10.1051/agro:2008062. Accessed 13 Jan 2024.

De Lima Caldas, E., & Jayo, M. (2019). Agriculturas urbanas em São Paulo: histórico e tipología. *Confins*. Retrieved at: http://journals.openedition.org/confins/18639. Accessed 13 Jan 2024.

De Terssac, G. (2012). La théorie de la régulation sociale: repères introductifs. *Revue interventions économiques, 45*. Retrieved at: https://journals.openedition.org/interventionseconomiques/1476#quotation. Accessed 13 Jan 2024.

Delisle, H. (1998). La sécurité alimentaire, ses liens avec la nutrition et la santé. *Revue canadienne d'études du développement, 19*(4), 307–329. Retrieved at: https://www.researchgate.net/profile/Helene-Delisle/publication/233258009_La_securite_alimentaire_ses_liens_avec_la_nutrition_et_la_sante/links/5a562124a6fdcc30f86d283c/La-securite-alimentaire-ses-liens-avec-la-nutrition-et-la-sante.pdf. Accessed 13 Jan 2024.

Di Fiore, G., Specht, K., Rover, O. J., & Cesare, Z. (2022). Stakeholders' social acceptance of e anew organic waste management policy in the city of Florianopolis (Brazil). *Journal of Cleaner Production, 379*. Retrieved at: https://doi.org/10.1016/j.jclepro.2022.134756. Accessed 13 Jan 2024.

Dubbeling, M., De Zeeuw, H., & Van Veenhuizen, R. (2011). *Cities, poverty and food; multistakeholder policy and planning in urban agriculture*. Royaume-Uni, RUAF Foundation & Practical Action Publishing Ltd. Retrieved at: https://ruaf.org/assets/2019/11/Cities-Poverty-and-Food.pdf. Accessed 13 Jan 2024.

Dubbeling, M., Van Veenhuizen, R., & Halliday, J. (2019). L'agriculture comme stratégie de réduction des risques face au changement climatique et aux catastrophes. In *Facts Report* (20), *L'agriculture urbaine, nourrir les villes autrement* (pp. 32–39). Retrieved at: https://www.institut.veolia.org/sites/g/files/dvc2551/files/document/2019/11/Agriculture%20urbaine%2C%20une%20revue%20de%20l%27Institut%20Veolia.pdf. Accessed 13 Jan 2024.

Duchemin, É. (Ed.). (2013). *Agriculture urbaine: Aménager et nourrir la ville*. VertigO. Retrieved at: http://www.editionsvertigo.org/images/livres/extraits/14.pdf. Accessed 13 Jan 2024.

Dumas, R-M. (2018). *Les enjeux politiques de la gouvernance environnementale : le cas de la gestion des eaux de surface à São Paulo. Québec, Université Laval, Department of Geography, unpublished MA dissertation.*

FAO. (2007). *Profitability and sustainability of urban and peri-urban agriculture*. Retrieved at: http://www.fao.org/3/a-a1471e.pdf. Accessed 13 Jan 2024.

FAO. (2010). *A new deal for school gardens. Promoting lifelong healthy eating habits*. Retrieved at: https://www.fao.org/3/i1689e/i1689e00.pdf. Accessed 13 Jan 2024.

FAO. (2014). *Growing greener cities in Latin America and the Caribbean*. Retrieved at: https://www.fao.org/3/i3696e/i3696e.pdf. Accessed 13 Jan 2024.

Giacchè, G. (2016). De la ville qui mange à la ville qui produit: l'exemple des Hortelões Urbanos de São Paulo. In *Espaces et sociétés* (40), Transition sociale et environnementale des systèmes agricoles et agro-alimentaires au Brésil. Travaux et documents de l'unité mixte de recherche 6590 (pp. 99–110). Retrieved at: https://hal.science/hal-01483364/document. Accessed 13 Jan 2024.

Giacchè, G. et Porto, L. (2015) "Politicas publicas de agricultura urbana e periurbana : uma comparação entre os casos de São Paulo e Campinas". *Informações Economicas, SP*, vol. 45, n° 6, p. 45-60. Retrieved at: http://www.iea.sp.gov.br/ftpiea/publicacoes/ie/2015/ie-1215.pdf. Accessed 13 Jan 2024.

Gravel, N. (2009, reedited in 2010). *Géographie de l'Amérique latine. Une culture de l'incertitude*. Presses de l'Université du Québec (Col. « Géographie contemporaine »), 372 p. Retrieved at: https://www.puq.ca/catalogue/livres/geographie-amerique-latine-1547.html. Accessed 13 Jan 2024.

Gravel, N. (2017). *Conflits et régulation dans l'application des politiques publiques d'agriculture urbaine au Brésil*. Paper presented as part of the colloquium "*L'agriculture urbaine comme stratégie de revendication et d'occupation des espaces verts en milieu urbain et périurbain*" at the 85th Congress of the Association francophone pour le savoir (ACFAS), McGill University, May 11–12.

Gravel, N., & Mestiri, R. (2015). *Le jardin communautaire comme pôle de gouvernance environnementale en milieu périurbain havanais, Cuba*. Presentation at the 83rd congress of ACFAS, Rimouski, Quebec, May 25–30.

Gravel, N., Dumas, R.-M., De Luca, E., Castanho, R., & Spinola, A. L. (2016). *La gobernanza de agua en la Región Metropolitana de São Paulo: El desafío de la integración de los eco-saberes en las políticas públicas en un contexto de cambios climáticos*. Paper presented at the First Latin American Ecosystem Partnership Conference, Cali, Colombia, October. 18–21.

Gravel, N., & Lemieux, J. (2022). L'agriculture urbaine et périurbaine au Brésil : entre activisme et institutionnalisation. Les cas des métropoles de São Paulo, Salvador de Bahia et de Florianópolis. *Cahiers de géographie du Québec, 67(187), 37–54*. https://doi.org/10.7202/1112468ar

Hall, A. (2006). "From Fome Zero to Bolsa Família: Social politics and poverty alleviation under lula". *Journal of Latin American Studies*, vol. 38, n° 4, p. 689-709. Retrieved at: https://eprints.lse.ac.uk/21244/1/From%20Fome%20Zero%20to%20Bolsa%20Familia%20%28LSERO%29.pdf. Accessed 13 Jan 2024.

Hovorka, A. J. (2005). "Gender, commercial urban agriculture and urban food supply" in Greater Gaborone, Botswana. In L. J. A. Mougeot (Ed.), *Agropolis* (pp. 137–152). IDRC. Retrieved at: https://idrc-crdi.ca/en/book/agropolis-social-political-and-environmental-dimensions-urban-agriculture. Accessed 13 Jan 2024.

Hunter, D., & Lightner, J. (2016). *Mason Bee Revolution. How the Hardest Working Bee Can Save the World One Backyard at a Time. Seattle, Washington, Skipstone.*

Jonsson, P. (2004) Vegetation as an urban climate control in the subtropical city of Gaborone, Botswana. *International Journal of Climatology*, vol. 24, n° 10, p. 1307-1322. Retrieved at: http://onlinelibrary.wiley.com/doi/10.1002/joc.1064/pdf. Accessed 13 Jan 2024.

Kalfagianni, A., & Skordili, S. (Eds.). (2018). *Localizing global food. Short food supply chains as responses to agri-food system challenges*. Routledge, 200 p.

Kulak, M., Graves, A., & Chatterton, J. (2013). Reducing greenhouse gas emissions with urban agriculture: A life cycle assessment perspective. *Landscape and Urban Planning, 111*, 68–78. Retrieved at: https://www.sciencedirect.com/science/article/abs/pii/S0169204612003209. Accessed 13 Jan 2024.

Le Velly, R. and Bréchet, J.-P. (2011) "Le marché comme rencontre d'activités de régulation: initiatives et innovations dans l'approvisionnement bio et local de la restauration collective". *Sociologie du travail*, vol. 53, n° 4, p. 478-492. Retrieved at: https://www.sciencedirect.com/science/article/abs/pii/S0038029611000951. Accessed 13 Jan 2024.

Millennium Ecosystem Assessment (MEA). (2005). *Ecosystems and human well-being: synthesis*. Island Press. Retrieved at: https://www.millenniumassessment.org/documents/document.356.aspx.pdf. Accessed 13 Jan 2024.

Mok, H.-F., Williamson, V. W., Groves, J. R., Burry, K., Barker, S. F., & Hamilton, H. J. (2013). Strawberry fields forever? A review of urban agriculture in developed countries. *Agronomy for Sustainable Development, 34*, 21–43. Retrieved at: https://research.monash.edu/en/publications/strawberry-fields-forever-urban-agriculture-in-developed-countrie. Accessed 13 Jan 2024.

Mougeot, L. J. A. (2000). *Urban agriculture: Definition, presence, potentials and risks, and policy challenges* (Cities Feeding People Series. Report 31). Retrieved at: https://idl-bnc-idrc.dspacedirect.org/server/api/core/bitstreams/a0cf4b0d-b96c-4124-a1de-f006d4a97f00/content. Accessed 13 Jan 2024.

Mougeot, L. J. A. (Ed.). (2005). *Agropolis. The social, political and environmental dimensions of urban agriculture*. IDRC. Retrieved at: https://idrc-crdi.ca/en/book/agropolis-social-political-and-environmental-dimensions-urban-agriculture. Accessed 13 Jan 2024.

Mougeot, L. J. A. (2005a). Introduction. In *Agropolis. The social, political and environmental dimensions of urban agriculture* (pp. 1–29). IDRC. Retrieved at: https://idrc-crdi.ca/en/book/agropolis-social-political-and-environmental-dimensions-urban-agriculture. Accessed 13 Jan 2024.

Mougeot, L. J. A. (2005b). Neglected issues on form and substance of research on urban agriculture. In *Agropolis. The social, political and environmental dimensions of urban agriculture* (pp. 267–279). IDRC. Retrieved at: https://idrc-crdi.ca/en/book/agropolis-social-political-and-environmental-dimensions-urban-agriculture. Accessed 13 Jan 2024.

Nagib, G. (2016). *Agricultura urbana como ativismo na cidade de São Paulo: o caso da Horta das Corujas*. Master's dissertation, department of Geography, University of São Paulo, Brazil. Retrieved at: http://www.teses.usp.br/teses/disponiveis/8/8136/tde-18082016-124530/en.php. Accessed 13 Jan 2024.

Nunez Sanchez, A. R., Viana Da Cruz, T., Narezi Gnarezi, G., & Silva Crepadi, M.-O. (2021). Organic production in the state of Bahia, Brazil: A spatio-temporal analysis of registrations and certification bodies (2014–2020). *Sociedade e Natureza, 33*(e60426). Retrieved at: https://www.redalyc.org/journal/3213/321366102031/html/. Accessed 13 Jan 2024.

Pereira, S. (2019). *Projeto de São Bernardo produz de mais de 1 tonelada de hortifrutis orgânicos*. Cidade de São Bernardo. https://www.saobernardo.sp.gov.br/maximizada/-/asset_publisher/5cLluTMVcxDN/content/projeto-de-agricultura-urbana-de-sao-bernardo-produz-de-mais-de-1-tonelada-de-hortifrutis-organicos-ao-mes?inheritRedirect=false. Accessed 13 Jan 2024.

Perlman, J. E. (2010). *Favela. Four decades of living on the edge in Rio*. Oxford University Press.

Petrini, C. (2007). *Slow food nation: A blueprint for changing the way we eat*. Rizzoli Ex Libris.

Prizendt, S. (2017). Urban Agroecology Movement (MUDA) changes the Brazilian metropolis. *RUAF Urban agriculture magazine, 33*, 68–69. Retrieved at: https://edepot.wur.nl/448793. Accessed 13 Jan 2024.

Reynaud, J. D. (1997). Les règles du jeu. L'action collective et la régulation sociale. Paris, A. Colin.

Reynaud, J. D. (1999). Le conflit, la négociation et la règle. Toulouse, Octarès.

Reyburn, S. (2002). "Le cadre de vie et les jardins potagers communautaires à Montréal". *VertigO la revue électronique en sciences de l'environnement*, vol. 3, n° 2. Retrieved at: http://journals.openedition.org/vertigo/3794. Accessed 13 Jan 2024.

Rivière D'Arc, H. (2006). *Le pari urbain en Amérique latine. Vivre au centre des villes*. L'Harmattan.

Rocha, A. A., & Feldmann, F. (2006). *A Mata Atlântica é aqui. E daí? História e luta da Fundação SOS Mata Atlântica*. Terra Viagem.

Rodrigues, E. A., Braga Moraes, R. A. V., Camacho Pires, B. C., & De Luca, E. F. (Eds.). (2020). *Serviços ecossistêmicos e bem-estar humano na reserva da biosfera do Cinturão verde da cidade de São Paulo*. Instituto Florestal. Retrieved at: https://semil.sp.gov.br/educacaoambiental/prateleira-ambiental/servicos-ecossistemicos-e-bem-estar-humano-na-reserva-da-biosfera-do-cinturao-verde-da-cidade-de-sao-paulo/. Accessed 13 Jan 2024.

Santandreu, A., & Lovo, I. C. (2007). *Panorama da agricultura urbana e periurbana no Brasil e diretrizes políticas para sua promoção. Identificação e caracterização de iniciativas de AUP em regiões metropolitanas urbanas*. Belo Horizonte, Minas Gerais, Ministério do Desenvolvimento Social e Combate à Fome and FAO. Retrieved at: http://www.agriculturaurbana.org.br/textos/panorama_AUP.pdf. Accessed 13 Jan 2024.

Shappo, S. (2021). Fome e insegurança alimentar em tempos de pandemia da covid-19. *SER Social, 23*(48), 28–52. Retrieved at: https://periodicos.unb.br/index.php/SER_Social/article/view/32423/28783. Accessed 13Jan 2024.

Sutton, S. (2021). Moringa Oleifera: The Miracle tree? *Renature*. July 19. Retrieved at: https://www.renature.co/articles/moringa-oleifera-the-miracle-tree/#:~:text=Mainly%2C%20it%20is%20because%20every,%2C%20medicine%2C%20and%20water%20purification. Accessed 13 Jan 2024.

Taguchi, M., & Santini, G. (2019). L'agriculture urbaine dans les pays du Nord et du Sud : une perspective de la FAO. In *Facts Report* (20) *L'agriculture urbaine, nourrir les villes autrement,* p. 12-17. Retrieved at: https://www.institut.veolia.org/sites/g/files/dvc2551/files/document/2019/11/Agriculture%20urbaine%2C%20une%20revue%20de%20l%27Institut%20Veolia.pdf. Accessed 13 Jan 2024.

Tallamy, D. W. (2018). *Bringing nature home. How you can sustain wildlife with native plants.* Timberpress.

Turgeon, J., & Savard, J.-F. (2012). Politique publique. *Le dictionnaire encyclopédique de l'administration publique.* Retrieved at: http://www.dictionnaire.enap.ca/fr/accueil.aspx?sortcode=1. Accessed 13 Jan 2024.

Visoni, C., & Nagib, G. (2019). La réappropriation de l'espace urbain via les jardins partagés au Brésil. In *Facts Report* (20). *L'agriculture urbaine, nourrir les villes autrement* (pp. 88–91). Retrieved at: https://www.institut.veolia.org/sites/g/files/dvc2551/files/document/2019/11/Agriculture%20urbaine%2C%20une%20revue%20de%20l%27Institut%20Veolia.pdf. Accessed 13 Jan 2024.

Visoni, C. (2024). Horta das Corujas. Minha historia de um pequeno paraiso em Sao Paulo e guia para plantio urbano. Sao Paulo: Bambual ed.

Wamsler, C. (2015) "Mainstreaming ecosystem-based adaptation: Transformation toward sustainability in urban governance and planning". Ecology and Society, vol. 20, n° 2. Retrieved at: https://www.ecologyandsociety.org/vol20/iss2/art30/. Accessed 13 Jan 2024.

Wamsler, C., Brink, E., & Rivera, C. (2013). Planning for climate change in urban areas: From theory to practice. *Journal of Cleaner Production, 50*, 68–81. Retrieved at: https://doi.org/10.1016/j.jclepro.2012.12.008. Accessed 13 Jan 2024.

Wegmuller, F., & Duchemin, É. (2010). Multifonctionnalité de l'agriculture urbaine à Montréal : étude des discours au sein du programme des jardins communautaires. *VertigO-La revue électronique en sciences de l'environnement, 10*(2). Retrieved at: https://doi.org/10.4000/vertigo.10445. Accessed 13 Jan 2024.

Websites

City of São Paulo. (2021). *Programa de metas 2021–2024.* São Paulo City Hall. https://www.prefeitura.sp.gov.br/cidade/secretarias/upload/governo/arquivos/programa_de_metas/programa-de-metas-2021-2024/programa_de_metas_2021-2024_versao-inicial.pdf. Accessed 13 Jan 2024.

City of São Paulo. (2022). *Homenagem ao Ambientalista Sergio Shigeeda* (in memoriam). https://www.prefeitura.sp.gov.br/cidade/secretarias/meio_ambiente/umapaz/formacao_em_educacao_ambiental/programacao_mensal/index.php?p=346006. Accessed 13 Jan 2024.

Forbes Agro Brasil. (2022). *No aniversário de São Paulo, suas hortas urbanas mostram o campo na cidade.* https://forbes.com.br/forbesagro/2022/01/no-aniversario-de-sao-paulo-suas-hortas-urbanas-mostram-o-campo-na-cidade/. Accessed 13 Jan 2024.

Government of Brazil, Chamber of Deputies. (2015). *Proposta cria a politica nacional de agricultura urbana.* https://www.camara.leg.br/noticias/465228-proposta-cria-a-politica-nacional-de-agricultura-urbana/. Accessed 13 Jan 2024.

Government of Brazil, Chamber of Deputies. (2017). *Camara aprova Politica Nacional de Agricultura Urbana.* https://www2.camara.leg.br/camaranoticias/noticias/AGROPECUARIA/548085-CAMARA-APROVA-POLITICA-NACIONAL-DE-AGRICULTURA-URBANA.html. Accessed 13 Jan 2024.

Milan Urban Food Policy Pact. https://www.milanurbanfoodpolicypact.org/#. Accessed 13 Jan 2024.

Municipal Market of São Paulo (*Mercado municipal de São Paulo*). (n.d.). *Tudo Sobre o Mercado de São Paulo.* https://www.mercadomunicipalsp.com/sobre-o-mercadao-de-sp/. Accessed 13 Jan 2024.

Municipality of Florianopolis (Prefeitura de Florianópolis). (2019). *Florianópolis cultiva mais de 100 hortas orgânicas e contribui para o bem-estar na cidade*. https://www.nsctotal.com.br/noticias/florianópolis-cultiva-mais-de-100-hortas-organicas-e-contribui-para-o-bem-estar-na-cidade. Accessed 13 Jan 2024.

Rede Semear Floripa website. https://www.visualagroecology.com/semear. Accessed 13 Jan 2024.

Slow Food. (2014). *A thousand gardens in Africa: The story so far.* https://www.slowfood.com/a--thousand-gardens-in-africa-the-story-so-far/. Accessed 13 Jan 2024.

Trading Economics. (2022). *Brazil—Urban Population*. https://tradingeconomics.com/brazil/urban-population-percent-of-total-wb-data.html#:~:text=Urban%20population%20(%25%20of%20total%20population)%20in%20Brazil%20was%20reported,compiled%20from%20officially%20recognized%20sources. Accessed 13 Jan 2024.

Chapter 9
Turning the Wheel of Social Innovation: Unveiling Water Governance Strategies in San Gabriel Chilac, México

Griselda Tomé Hernández, Nathalie Gravel, and Sonia Emilia Silva Gómez

Abstract The San Gabriel Chilac case illustrates community-driven innovation in managing drinking water. Leveraging local topography, the community implemented a gravity-driven distribution system, reflecting sustainable practices from precolonial eras. The network-type management, facilitated through participatory mechanisms, emphasizes interaction and dialogue. Decision-making initiates in community assemblies, promoting bidirectional communication. This model empowers local actors, fostering active participation, streamlined decision-making, and issue resolution, enhancing community trust. Recognizing and strengthening local capacities is crucial for effective water governance, considering the context dependency of policies. The region's water management reflects a historical blend of pre-Columbian and Spanish influences, underscoring the need to comprehend these local contexts for informed decision-making and future initiatives.

The main author of this chapter is Dr. Griselda Tomé Hernández, professor-researcher at the Instituto Tecnológico Superior de San Martín Texmelucan (ITSSMT). This case study derives from her doctoral research under my supervision at Université Laval, Canada.
Dr. Sonia Emilia Silva Gómez is Professor researcher at the Department of Sustainable Development, Scientific Institute of the Benemérita Universidad Autónoma de Puebla (BUAP).

G. Tomé Hernández
Instituto Tecnológico Superior de San Martín Texmelucan (ITSSMT),
San Martín Texmelucan de Labastida, Mexico

N. Gravel (✉)
Department of Geography, Université Laval, Quebec City, QC, Canada
e-mail: Nathalie.gravel@ggr.ulaval.ca

S. E. S. Gómez
Department of Sustainable Development, Scientific Institute of the Benemérita Universidad Autónoma de Puebla (BUAP), Puebla, Mexico

N. Gravel, *Tending Nature*, https://doi.org/10.1007/978-3-030-47062-3_9

Keywords Agricultural intensification · Decision-making processes · Drinking water · Filtrating galleries · Groundwater · Irrigation systems · Mexican agricultural history · Network-type management · Participatory water governance · San Gabriel Chilac · Social innovation · Tehuacán Valley · Water · Water distribution system

9.1 Introduction

For over 20 years now, humanity has been grappling with a severe water crisis, which fundamentally stems from a governance crisis due to the inadequate management of water resources (UNESCO, 2003, 2023). Mexico is not exempt from this challenging water management landscape, as evidenced by the extensive, intricate, and varied water issues faced by its 31 states and Mexico City. Despite the considerable changes and progress made by the National Water Commission (CONAGUA), particularly in establishing a system for social participation in decision-making concerning water management in 1992, the Mexican water sector still faces the formidable task of achieving sustainable, participatory, and integrated water resource management. This challenge is underscored by case studies conducted by Tomé Hernández (2019) in the Salado River subbasin, Soares et al. (2009) in the Amacuzac River basin, Kauffer Michel (2005) in the Usumacinta and Grijalva Rivers, Kauffer Michel (2008) on watershed committees in Chiapas and Tabasco, Cisneros (2008) in Laguna Seca (Guanajuato), Comundú (Baja California Sur), Casas Grandes (Chihuahua), and San Juan del Río (Querétaro), Vera Carta (2005) on the coast of Chiapas, Caire (2004) in the Lerma-Chapala watershed, and Vargas Valencia et al. (2004) in the Lake Patzcuaro Basin.

Enhancing processes in water resource management often involves not just technical aspects but a fundamental shift in perspectives on water. Primarily, restructuring power dynamics to accommodate the needs of all water users, including Nature, and promoting their active participation in decision-making processes form the foundation for effective governance of common resources (Brooks et al., 2009; Gleick, 2002; Ostrom, 2010).

While this holds true for the management of irrigation water in a scarcity context, as exemplified by Elinor Ostrom's case studies in Spain, it is crucial to acknowledge endeavors focused on building local capacities for managing drinking water supply systems (Pimentel-Equihua et al., 2012). Additionally, efforts to formulate strategic proposals that consider the specific potentials, and constraints of each territory, as well as the socio-economic marginalization of particular groups, contribute to more tailored, just and effective approaches (Schejtman & Berdegué, 2003, cited in Schneider & Peyrè Tartaruga, 2006).

The present case study conducted by author Griselda Tomé Hernández for the completion of her Ph.D. dissertation under my supervision and coauthor Dr. Sonia Emilia Silva Gómez in the subbasin of the Salado River in the state of Puebla, Mexico, is notable for its unique exploration of collective drinking water management, filling a gap in existing literature. Pioneering researchers like Ostrom, who have significantly shaped economic governance frameworks and water rights practices, have primarily focused on irrigation water for agricultural purposes, neglecting drinking water management. This contribution stands out as it sheds light on a water management system inspired by the Castilian turn system, providing users with considerable flexibility and autonomy in allocating scarce water resources among landholders (Ostrom, 2010). The study addresses a question raised by Ostrom herself regarding the potential outcomes for Latin America if the Aragon water distribution model had been implemented instead of the Castilian one, both originating from Spain, the birthplace of Ferdinand and Isabella, monarchs of Spain who colonized the territory of New Spain. According to Ostrom (2010), the main differences between the Aragon and Castilian water distribution models lie in their approaches to management and decision-making. The Aragon model emphasizes community involvement, collective decision-making, and proportional allocation, fostering a cooperative environment. In contrast, the Castilian model is more centralized, with fixed allocations and individual ownership of water rights, which can lead to more hierarchical and rigid management structures. The case of San Gabriel Chilac is a testament to local communities addressing the long-standing issue of inequitable water access.

The primary objective of this study is to unveil a case of social innovation by providing a comprehensive exploration of the development, design, and implementation of drinking water management in San Gabriel Chilac, with a particular focus on the organizational structure of this model. This analysis of water management processes is guided by the theoretical principles of participatory water resources governance. The study employs a qualitative data collection and analysis methodology, incorporating field observations and surveys conducted with water rights holders and key informants.

The chapter's structure comprises five sections. The first section outlines the physiographic characteristics of the studied territory. The second section delves into the definition, dimensions, and principles of the concept of environmental governance. Following this, the theoretical foundations of participatory water governance are applied to analyze a real-world situation, specifically water allocation for drinking purposes in the town of San Gabriel Chilac.

9.2 Study Site

The study was conducted in the municipal seat of San Gabriel Chilac, situated in the southeastern part of the state of Puebla, Mexico, approximately 74.5 miles from the capital city of the same name. Historically, it was part of the ancient district of Tehuacán, inhabited by an Aztec tribe engaged in corn cultivation in the Tehuacán valley. Another tribe, belonging to the Tolteca group and governed by a woman—who was the chief's spouse—resided in the area, engaging in trade of chiles with the first tribe.

Upon the Spanish conquest, the district was named Tenochtitla, signifying the union of tribes, with the "señora de los chiles" (the "chile lady") governing the territory. San Gabriel Chilac is one of the 24 municipalities within the Salado River subwatershed, positioned downstream of the Salado River, near its confluence with the Papaloapan River.

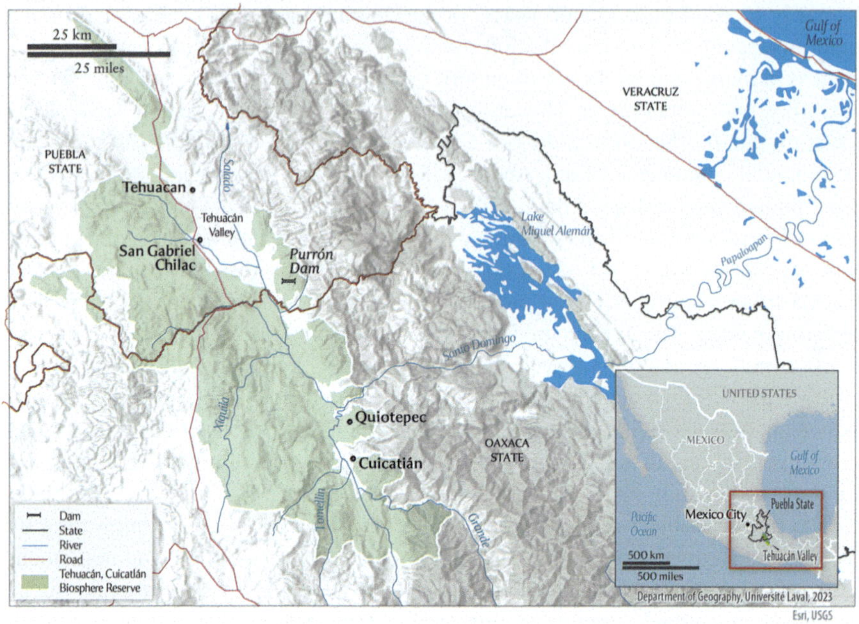

Fig. 9.1 Map of the Tehuacán Valley

The Rio Salado (Salted River) flows north–south through the Tehuacán Valley, serving as a major tributary to the Papaloapan River, which eventually empties into the Gulf of Mexico near the port of Alvarado in the state of Veracruz. The name "Salado" originates from the significant salt (NaCl) content it carries, particularly after merging with the Tehuacán river that traverses the Zapotitlán depressions. The Tehuacán region has been renowned for its salt mines since its early settlement, playing a central role in the development of crucial regional trade (Roulet, 2015). The Spanish interest in salt was evident in their efforts to pacify the Tehuacán region during indigenous resistance, notably during the conflict between the Franciscan and Dominican orders in 1570 (Ibid).

The municipality of San Gabriel Chilac spans an area of 26,998.69 acres (109.26 km)[2] (see Fig. 9.1) and is home to a population of 15,954 inhabitants, primarily of indigenous origins, with a mix of Tolteca and Azteca descent (INEGI, 2020). Socioeconomic indicators reveal a moderate poverty rate, with 61% experiencing moderate poverty and 23.6% facing severe poverty (Government of Mexico, 2022). The Marginalization index, calculated by CONAPO (Consejo Nacional de Población) (2020), indicates a low level of marginalization for both the municipality of San Gabriel Chilac and its eponymous locality. However, outside the locality of San Gabriel, a medium level of marginalization is observed. The rural areas exhibit varying degrees of social marginalization, ranging from moderate to acute.

This socio-economic marginalization is concretely manifested at the household level. For instance, 0.59% (25 houses) lack connection to the drinking water system in the municipality of San Gabriel Chilac out of a total of 3897 households.

Fig. 9.2 Corn fields in the Tehuacán Valley. (Credit: Gravel, 2015)
Description: The cultivation of corn, a staple crop with deep cultural roots, has been a fundamental practice in the Tehuacán Valley region for centuries. This crop not only contributes to the sustenance of local communities but also serves as a testament to the resilience and adaptability of agricultural practices in the semiarid environment of the Tehuacán Valley. The interplay between natural resources, climate, and human ingenuity is evident in the thriving corn fields that shape the agricultural landscape of the Tehuacán Valley

Additionally, 4.5% of these households do not have access to electricity. The limited access to basic services in a small percentage of households is often attributed to the recentness of construction and the challenging terrain in remote areas. This, in turn, affects access to education and social services for residents. Notably, houses built on the foothills of the mountains due to urban sprawl face difficulties in accessing services (Government of the state of Puebla, 2014).

The municipality is situated near the Biosphere Reserve of Tehuacán-Cuicatlán and is home to the Water Museum (Museo del Agua), inaugurated in 2005 and funded and operated privately. To the north of the valley lies the pristine Tehuacán aquifer, a source of the naturally mineral-rich Tehuacán water. The semiarid Tehuacán-Cuicatlán Valley has a rich history dating back to pre-Columbian times, characterized by extensive irrigated agriculture that yields significant corn and fresh produce harvests. The valley boasts abundant biodiversity, contributing to its overall ecological richness (Woodbury & Neely, 1972; Enge & Whiteford, 1989; Neely et al., 2015; Vallejo et al., 2019; Tomé Hernández, 2019) (Figs. 9.2 and 9.3).

Fig. 9.3 View from the Tehuacán-Cuicatlán Valley Biosphere Reserve with saguaro cacti and the Sierra Madre del Sur Mountains in the background. (Credit: Gravel, 2015)
Description: The view from the Tehuacán-Cuicatlán Valley Biosphere Reserve is a breathtaking panorama that captures the essence of its unique ecosystem. Saguaro cacti, with their towering and distinctive forms, stand as iconic features in the foreground, symbolizing the arid yet resilient nature of the region. In the background, the Sierra Madre del Sur Mountains add to the scenic beauty of the landscape

The region supports a diverse range of ecosystems influenced by varying altitudes and was granted recognition as a World Heritage site by UNESCO in 2018 (Vallejo et al., 2019). The Tehuacán-Cuicatlán Valley is positioned between two north–south mountain ranges, with the eastern range intercepting moisture from the Gulf of Mexico. Consequently, the valley is situated in a rain shadow, resulting in its semiarid climate. Characterized by a rainy season spanning from June to September and an average annual precipitation ranging between 9 and 20 inches (250–500 mm), effective water management has played a crucial role in the settlement and development of agricultural activities in the region. This is evident in the presence of prehistoric irrigation systems (Woodbury & Neely, 1972).

San Gabriel Chilac exhibits mountainous relief to the west, marked by the foothills of the Sierra de Zapotitlán. The urban sprawl is situated on flat terrain, featuring a gentle slope in a northwest–southeast direction and varying in elevation from approximately 3770 to 5580 feet (1150–1700 meters) above sea level. The prevailing climate is semiarid, characterized by an average annual temperature ranging between 57 °F and 75 °F (14 °C and 24 °C), with an average annual rainfall of approximately 13 inches (350 mm) (lower than the national average of 30 inches (750 mm) per year).

Groundwater serves as the primary supply source for various consumptive uses, including agricultural, domestic, and livestock purposes in the territory. During the early colonial period, two water sources were identified, leading Franciscan friar Francisco Medina de Santa Cruz to guide the community in constructing a channel that bifurcated to irrigate lands below. Friar Medina was part of the Spanish conquerors dispatched to convert and take possession of the lands of New Spain starting in 1529 (Roulet, 2015).

To maintain water infrastructures, a collaborative system was instituted (Gil Huerta, 1972), involving collective cleaning chores scheduled every Monday—a practice that has endured unchanged in the community to this day. The same friar oversaw the construction of the Chapel of San Gabriel Chilac between 1541 and 1543, built by the Catholic converted tribes (Idem). Following its completion, the parish was established, named after the interceding Archangel Gabriel, as suggested by the Lady of the chiles (Idem). In 1543, an official ceremony conferred legal ownership of the water sources to the municipality, adorned with the colors and arms of Castilla and Leon under the authority of King Carlos V of Spain (Idem). This historical background is crucial for understanding the evolving water control dynamics in the area.

9.3 Methodological Considerations

This chapter was crafted through a mixed-methods approach that involved field observations, semi-structured interviews conducted in 2015 and 2016, oral testimonies, and data collection from second-hand sources. A total of 60 semi-structured interviews were undertaken between February and August 2016 by the main author of this chapter, Griselda Tomé Hernández, comprising 53 with users and 7 with key informants such as presidents of community committees and community leaders.

For the quantification and analysis of the data derived from the questionnaires, the SPSS® program version 11.5 (Statistical Package for Social Sciences) was utilized. This rigorous methodology ensured a robust and comprehensive exploration of the subject matter, combining qualitative insights from interviews with quantitative data analysis for a holistic understanding of the water management dynamics in the region.

9.4 The Tehuacán Valley: Pre-Colombian Roots of Hydraulic Societies

The Tehuacán Valley stands as a testament to the ancient origins of irrigated agriculture in Mesoamerica, with evidence pointing to agricultural intensification as early as 700 B.C. (Neely et al., 2015: 347). A pivotal development was the construction of the monumental Purrón Dam Complex in the state of Puebla, dating back to 750–600 B.C. This complex, built over four different periods and levels, marked a transformative era when human settlements were still small and decentralized (Woodbury & Neely, 1972: 92, 88).

The introduction of irrigation through channeling water enabled the cultivation of a second annual corn crop, expanding agricultural possibilities. Additionally, the dam's construction facilitated the introduction of new plants like cotton, black and white sapote, and avocado, which were not native to the region. The water control and irrigation features became invaluable assets for local farming communities in the Rio Salado region. This led to the formation of 19 villages around the Purrón Dam Complex, enhancing access to food supplies for the local population. The construction of the dams required significant manpower, with the fourth dam at Purrón estimated to have involved around 1000 workers (Idem).

Archaeological evidence, including spillways discovered by the Tehuacán Archaeological-Botanical Project in 1961, showcased advanced water management techniques to channel water to nearby fields. Despite challenges such as erosion, silting, dam breaching, and dense vegetation cover, this ancient complex remains 90% intact, reflecting its resilience over time (Woodbury & Neely, 1972: 97).

The Purrón Complex's set of four dams faced abandonment due to silting, rendering them ineffective. Before integration into the Aztec Teotihuacán-based regime (present-day Mexico City), pre-Colombian independent settlements in the Tehuacán Valley had already designed irrigation systems. The abandonment of the dams coincided with evidence of a population decline in the region (Idem).

Following this period, the remaining population experienced Spanish conquest, and the encomienda and hacienda systems were implemented. This system, relying on different water distribution techniques, granted control of water distribution to the landowner, as per the Castilian system (Neely et al., 2015; Hernández Garciadiego & Herrerías Guerra, 2004).

Following the Conquest, the social organization that emerged to manage water allocation, specific to each local territory and tailored to the needs of water users, might have been influenced by Spanish friars. These friars sought fair and equitable solutions to water provisioning in the territories under their oversight. Given the scarcity of missionary labor and the expansive territory of New Spain, each residing friar likely had a significant geographical influence over local managerial decisions (Neely et al., 2015). The friars, alongside the overseer from the Audiencia, emerged as the primary Spanish territorial managers, collaborating with local tribal chiefs. It is crucial to elucidate the roles of these historical actors and the legacy passed down to present inhabitants. This historical contextualization provides depth to the study and enhances our understanding of the evolving relationship between people, land, and water over time in the Tehuacán Valley.

9.5 Theoretical Foundation of Water Governance

While the term "governance" is ubiquitous in scientific literature, providing a precise definition proves challenging due to its myriad meanings and contradictions across various sciences and disciplines (Brody, 2009). Given that social participation forms the backbone of comprehending common resource governance in this study, we will adopt the definition put forth by the Commission on Global Governance:

> Governance is the sum of the many ways individuals and institutions, public and private, manage their common affairs. It is a continuing process through which conflicting or diverse interests may be accommodated and co-operative action may be taken. It includes formal institutions and regimes empowered to enforce compliance, as well as informal arrangements that people and institutions either have agreed to or perceive to be in their interest (UN Commission on Global Governance, 1995 in Gravel and Lavoie, 2009:17).

When applied to water, the concept of governance is often regarded as the most suitable theoretical approach for discussing effective participatory resource management with decentralized decision-making schemes at various scales (Infesta Dominguez et al., 2012). This is reflected in the following understanding of water governance:

> The range of political, organizational and administrative processes through which communities articulate their interests, their input is considered, decisions are made and implemented, and decision-makers are held accountable for developing, managing water resources and the delivery of water services (Bakker, 2003: 5).

These interpretations emphasize that environmental governance centers around a comprehensive social decision-making process (Brandes, 2005). Power is wielded for the administration of a common good (Barriga et al., 2007; Ostrom, 2010), demonstrating a clear emphasis on the relegitimization and modernization of public action. It involves the implementation of horizontal action mechanisms, a transition to more open forms of rationality, and a certain delegation of powers to civil society, decentralized collectivities, or autonomous institutions (national or international) (Theys, 2002; Gravel and Lavoie, 2009).

According to Gagnon (2000) and Moench et al. (2003, cited in Akhmouch, 2015), environmental governance transcends a series of hierarchical administrative actions in a specific sector and scale. It is more concerned with the process of decision-making—how, by whom, and under what conditions decisions are made—than the decisions themselves. This approach also considers the roles and responsibilities exercised by both formal and informal institutions.

Brody (2009) and Brandes (2005) argue that the concept of environmental governance enables the identification and description of autonomous networks, acknowledging their capacity to operate beyond government authority. Gaudin (2002, cited in Beaurain, 2002) asserts that public action within networks can be analyzed as a relational practice of cooperation that is not predetermined and is continually reinvented. Ostrom (2010) emphasizes that local or regional government should serve as a supporter and facilitator of social organizations, highlighting the importance of the autonomy of actors. This observation was evident in her research on water management in various regions of Spain, where she found that irrigators in certain regions had greater autonomy compared to those in other parts of the country. Ostrom's use of a case study in alpine pastoral communities in Switzerland, documented by Mc C. Netting (1981), also suggests that regions lying on the outskirts of state surveillance can become creative grounds for self-organizing the allocation of common resources (see Chap. 2). According to Ostrom (2010), users-owners constitute the main decision-making unit and act within a larger framework of rulemaking and governance that considers them legitimate. Their retention of local knowledge and a unique understanding of terrain characteristics, temporal variations, and the needs of their livestock make them the most suitable actors to establish a common set of rules and obligations.

Following Ostrom (2010: 250), the state should support social innovations by facilitating "collective access at low cost to information, transformation, and application," enabling appropriators of common goods to engage in a community-based system of water allocation.

Many authors argue that effective management and conservation of natural resources occur best at the local political level. The participation of local actors, closely connected with natural resources, is crucial for achieving protection objectives, contrasting with exclusionary conservation approaches (De Boer & Baquete, 1998; Heinen, 1996). This discussion prompts reflection on the state's role and the degree of decentralization needed in decision-making. Ostrom (2010) emphasizes the need for a regulatory framework that allows stakeholders to organize around their own rules within a legal structure legitimized by state institutions. The Mexican state's 1992 National Waters Law, recognizing citizens' participation and supporting local water user organizations, illustrates the state acting as a facilitator for collective water resource management (Pimentel-Equihua et al., 2012).

In this research on participatory water governance, the case study illustrates local, decentralized governance. According to Bagnasco and Le Gales (1997, cited in Gravel & Lavoie, 2009), local environmental governance involves "the power of a community to manage its natural resources within a territory appropriated to achieve collectively agreed-upon objectives and meet members' needs."

While the scientific literature presents various operational principles of environmental governance, this study follows those proposed by Akhmouch (2015):

(a) *Transversal Coordination*: Recognizing that water sector decisions are interconnected with other sectors, this dimension involves collaboration among Federal, State, and Municipal agencies to ensure policy coherence.

(b) *Participation Spaces for Dialogue*: Emphasizes the need to establish forums for diverse stakeholders to participate in decision-making, defining roles, responsibilities, and accountability.
(c) *Decision-Making*: Involves identifying and including various actors (public organizations, private sector, civil society, local population, indigenous communities, water users, and academics) in the decision-making process for water policies.

Similarly, Enjolras (2005), Gagnon (2000), and Brody (2009) underscore the significance of factoring in the characteristics of involved actors—such as objectives, needs, knowledge, values, local capacities, institutional forms, resources, and legitimacy—during decision-making. They emphasize addressing debates and divergent interests among actors to ensure decision-making takes place in a consensual manner.

(d) *Regulatory Frameworks*: These are essential for establishing operational social institutions (both formal and informal) in decision-making. They regulate and facilitate social relationships, guide interactions, and address conflicts or potentials.

Enjolras (2005) and Barriga et al. (2007) highlight that regulatory frameworks establish boundaries, agreements, roles, and responsibilities for various actors, emphasizing the importance of ensuring compliance with formal obligations—legal, political, hierarchical, or societal.

(e) *Accountability, Surveillance, Monitoring, and Follow-up*: This dimension emphasizes the importance of fulfilling formal obligations. Decision-makers should be accountable to stakeholders for their decisions and implementation, highlighting the need for responsible actions and decision-making.
(f) *Clear Assignment of Roles and Responsibilities*: Clearly defining and assigning roles and responsibilities to all actors in the water management process is crucial. This clarity prevents gaps and ensures the effectiveness of the water policy cycle. Mapping institutional roles and responsibilities aids in diagnosing inconsistencies.
(g) *Self-Financing Capacity*: Emphasizes the need for the water management entity to generate its own financial resources, ensuring it can fulfill its responsibilities in developing and managing water resources and delivering water services.
(h) *Training*: Improving water governance requires mechanisms and incentives to enhance training at all levels for fulfilling water-related responsibilities.
(i) *Availability of Policy-Relevant Data and Information*: Access to reliable, up-to-date hydrological, technical, social, economic, and financial data is crucial. This enables objective analysis and the development of effective water strategies at different levels.

Fig. 9.4 (**a** and **b**) Water Infrastructure in San Gabriel Chilac. (Credit: Griselda Tomé Hernández, 2016)
Description: Remnants of a water tank situated in the heart of San Gabriel Chilac

9.6 Background to the Branch Model

The locality of San Gabriel Chilac offers a fascinating context for the exploration of the evolving dynamics of participatory water governance, with its roots extending back to the early twentieth century, as elucidated by the accounts of key informants.

According to oral testimonies from key informants, the contemporary "branch model" for domestic water management has historical antecedents in the erstwhile "model of fountains or pools," a concept conceived by the residents of Chilac themselves. The water distribution system comprised a dam that stored water from the La

Taza spring and an open canal that transported the stored water to the designated "fountains or pools." Positioned at each intersection of the town's narrow streets were two pools: one for storing water meant for human consumption and the other for providing drinking water to animals such as donkeys, oxen, horses, and sheep (see Fig. 9.4a, b). By the 1930s, the small town of San Gabriel Chilac boasted a total of 20 fountains, water tanks, or pools.

The water supply system for human consumption operated through a committee comprising a general president, known as the "president of the 20 sources," and the 20 presidents of the sources, each overseeing a specific fountain. The general president coordinated the maintenance efforts, while the fountain presidents organized users, ensured pool cleanliness, maintained pipes, and contributed to community activities. This model persisted until 1973 when piped drinking water was introduced, leading to the gradual disappearance of most pools. Today, only one pool remains, serving as a poignant testament to this history.

For some time, the municipal government oversaw the drinking water system, but due to inefficiencies, the residents of Chilac collectively advocated for taking charge of its administration. In 1986, a revival of the forgotten fountain or pool model was proposed, now rebranded as branches. Each branch, comprising specific street blocks, was established as a subdivision of the owners' association for water management. This model introduced unique mechanisms and procedures, fostering local governance through collaborative networks of engaged stakeholders, as further elucidated in the subsequent section.

9.7 Evaluating Water Governance in the Chilac Branch Model

This section delves into the examination and exposition of the internal framework of the branch model, elucidating its functionality, the diverse actors comprising it along with their interrelationships, and the resources integral to this groundbreaking drinking water management model. The analysis will employ water governance indicators. Prior to this, a brief overview of the physical components, namely, water infrastructures, within the gravity-fed drinking water supply system will be provided.

The infrastructure for the gravity-fed water supply system in San Gabriel Chilac is a vital component ensuring the reliable delivery of water to the community. The key elements of this system include:

(a) *La Taza Spring*: This natural spring serves as the exclusive source of water for domestic use in the community. The quality and sustainability of this source are critical for the entire water supply system.
(b) *Reservoir*: The reservoir functions as a storage facility for the water sourced from La Taza. It allows for the accumulation and controlled release of water, contributing to the consistent availability of water for the community.

(c) *Closed Canals*: Two closed canals are employed to transport water from La Taza to strategically positioned storage tanks. These canals operate on the principle of gravity, facilitating the efficient movement of water through the system.
(d) *Storage Tanks*: Strategically located at the town's entrance on a steep slope, these storage tanks play a crucial role in ensuring a stable water supply. They receive water from the canals and act as distribution points for further conveyance to households.
(e) *Pipe Network*: The pipe network serves as the lifeline of the system, distributing water from the storage tanks to individual households. The extensive network comprises approximately 2600 water intakes, providing direct access to water for the community residents.

This well-designed and integrated infrastructure is central to the success of the gravity-fed water supply system, enabling the efficient management and equitable distribution of water resources in San Gabriel Chilac.

9.7.1 The Organizational Structure and Actors

The organizational structure of the branch model is made up of two management levels: a Central Committee (CC) and a Branch Committee (BC)—in total, there are 52 branch committees—with the CC being the central administrative structure in this model.

The organizational structure of the branch model in San Gabriel Chilac is a multitiered system designed for effective management and governance of the community's drinking water supply. This structure consists of two primary levels:

Central Committee (CC): At the top of the hierarchy is the Central Committee, serving as the central administrative structure for the entire model. The CC holds a pivotal role in coordinating and overseeing the various aspects of water management in the community. It plays a crucial role in decision-making, policy formulation, and ensuring the overall functionality of the water supply system.

Branch Committees (BC): The community is subdivided into 52 Branch Committees, each representing a specific area or sector within the locality. These committees form the decentralized units responsible for localized water management. The BCs play a vital role in implementing policies, addressing community-specific needs, and facilitating communication between the central administration (CC) and the residents.

This organizational framework establishes a clear chain of command and facilitates efficient communication and decision-making processes. The combination of a centralized oversight (CC) and decentralized branches (BC) ensures that water management strategies can be tailored to the unique requirements of different sections of the community, fostering a responsive and community-centric approach to governance.

9.7.2 *The Formation of the Central Committee*

The Central Committee (CC) is a crucial component of the branch model, and its composition and election process play a significant role in the overall governance structure. The CC consists of 11 members, each with specific roles and responsibilities. The composition of the CC is determined through an annual election process, which involves the following steps:

Representation from Branches: Initially, representatives from branches 2 to 12 are elected. These representatives form the pool of candidates for the CC.

CC Member Election The 11 representatives elected from the branches then attend the final Assembly of the Central Committee, where they are designated to specific roles within the CC. The roles include a president, a secretary, a treasurer, seven members, and a messenger.

Presidential Election To elect the central president, attendees propose three candidates. Each proposed candidate is voted on individually, and the one with the highest number of votes becomes the president of the CC.

Secretary and Treasurer Election The same process is followed to elect the secretary and treasurer. Three candidates for each position are proposed, and attendees vote to determine the individuals who will hold these roles.

Messenger and Vocal Election The messenger and the seven members of the CC are proposed directly by the attendees. Presumably, the messenger and members are selected based on their qualifications, dedication, and contributions to the community.

This democratic and participatory election process ensures that community members have a say in choosing their representatives, fostering transparency and accountability within the governance structure of the branch model.

9.7.3 *The Formation of the Branch Committee*

The Branch Committee (BC) plays a pivotal role in the branch model, representing the interests and needs of specific user groups within the community. The structure of the BC includes a president, a secretary, a treasurer, and sometimes one or two additional members, particularly in branches with more than 100 users.

The annual election of the BC board of directors follows a unique process that differs from the direct voting method used for the Central Committee (CC). The election is based on the order of the list of users, who are owners of the water intakes within the branch. The election process unfolds as follows:

List of Users A list of users (water intake owners) within the branch is established.

Selection Process The election involves selecting the first three or four users from the list, depending on the size of the branch. The first user becomes the president, the second user becomes the secretary, the third user becomes the treasurer, and the fourth and fifth users, if applicable, become additional members.

Subsequent Elections For the next election cycle, the counting starts from the second user on the list, who will now assume the position of president. The following users on the list are designated as secretary, treasurer, and additional members in a consecutive manner.

This election approach ensures a systematic rotation of roles within the BC, providing opportunities for different users to assume leadership positions over time. It reflects a commitment to inclusivity and a fair distribution of responsibilities within the community's water governance structure.

The election process for both the Central Committee (CC) and the Branch Committee (BC) underscores a commitment to gender inclusivity, emphasizing that women can hold positions in either committee as long as they are water intake owners. These positions are considered honorary and service-oriented, as committee members do not receive any salary for their roles.

A key figure in the water governance structure is the water carrier, who holds a unique and crucial role in the effective functioning of the system. The primary responsibility of the water carrier is to manage the opening and closing of flow valves, directing water to different branches based on established shifts. Notably, the water carrier is the only role within the management model that receives remuneration for their services. Due to their extensive experience and comprehensive knowledge of the pipeline network, the water carrier holds a long-standing position and is not replaced annually. This underscores the significance of their role in ensuring the smooth operation of the water distribution system.

In the governance structure of the water management model in San Gabriel Chilac, the only government entity involved is the City Council. However, it plays a noninterventionist role in the decision-making process. Its contributions are limited to providing chlorine for water disinfection and making annual payments of rights related to water extraction and usage to CONAGUA.

It's important to highlight that the organizational structure and election processes in the branch model of San Gabriel Chilac differ from other water management models discussed in the literature, such as those described by Pimentel-Equihua et al. (2012) and Sandoval Moreno and Günther (2015). In the models studied by these authors, there is a single central structure with 12 and 7 members, respectively. The replacement time of CC members and the honorary nature of the positions, however, align with the model referenced by Pimentel-Equihua et al. (2012).

9.8 Spaces for Dialogue and Decision-Making in Water Governance

This section delves into the critical aspects of spaces dedicated to dialogue and decision-making within the innovative water governance model in San Gabriel Chilac. It explores how various actors engage, participate, and contribute to the decision-making processes that shape the management of drinking water resources in the community.In the examined case studies by Aguilar et al. (2001), Pimentel-Equihua (2006), Pimentel-Equihua et al. (2012), and Sandoval Moreno and Günther (2015), alongside Tomé Hernández (2019), the predominant focus lies on the establishment of agreements and regulations concerning the utilization of natural resources within community or ejido settings, with the assembly being recognized as the highest authority. Aligning with this tradition, decision-making in the present study is also conducted within the Assembly framework. However, a notable distinction exists in this management model, where two types of Assemblies coexist: (1) the Assembly overseen by the CC, gathering the 11 members of the CC and the 52 presidents of the branches; and (2) the Assembly held in each branch, providing a forum for the users within the branch, along with the board of directors or committee of the branch.

The creation of this space for dialogue by local actors reflects a decision-making process characterized by both representativity and participation. It is representative in the sense that all users are involved, either directly or indirectly, in both types of assemblies. Direct participation occurs in the Branch Committee (BC) Assemblies, while indirect representation takes place in the Central Committee (CC) Assembly. In the latter, the presidents of the 52 branches serve as spokespersons for the users, acting as the bridge that connects both levels:

> In the CC Assemblies, the central president informs the presidents of each branch about the problems, the actions that must be carried out, the agreements, the plans, the annual report, etc. Subsequently, the president of the branch transmits this information to the users of the branch and vice versa (Translated from Spanish, Interview with Member A, 2016[1]).

The model is participatory because within this social space, the interests, disagreements, proposals, needs, and knowledge of users (both men and women) converge during decision-making processes related to the formulation, implementation, and monitoring of regulatory decisions.

> In my branch I proposed that they reuse the water with which clothes are washed and the people accepted it. Later, this practice was implemented in other branches. Knowledge is spread in other branches through communication between neighbors (Interview with Member B, 2016).

For decision-making, local actors have developed direct communication mechanisms both from the top down and vice versa. These communication mechanisms facilitate active participation by all members and the achievement of decisions

[1] All interviews were carried in the Spanish language for this section.

through consensus. Additionally, they contribute to the promotion of local democracy, enabling users and leaders to take on commitments, as illustrated in the following testimony:

> Once the agreements are proposed in the CC Assembly or any important issue is reported, the 52 presidents of the branches have the obligation to inform and agree on the agreements with the users of their branch. Subsequently, the branch president informs the CC Assembly of the final decision through an official letter. If more than 26 answers are obtained in favor, the agreement is implemented. We are governed by what the majority thinks (Interview with Member C, 2016).

Similarly, they facilitate the identification of problems and effective coordination to resolve them:

> In the event that the user does not have water, he has to report his problem to the president of the branch. Later, he will communicate it to the central president. The latter will take care of solving the problem. The problem is solved, maximum in three days (Interview with Member C, 2016).

> With this system, problems can be detected and fixed quickly. Thus, conflicts and tensions are not generated (Interview with Member C, 2016).

For the development of the assemblies, both the CC and the BC follow the following protocol: roll call (users and/or presidents of the branches); declaration of the existence of the legal quorum; reading the agenda; reading of the minutes of the preceding assembly; reading of the reports that the branch presidents deliver to the CC; and general business. The assembly concludes when the attendees sign the minutes of agreements.

9.9 Operationalizing the Framework: Defining Roles, Ensuring Accountability, and Overseeing Actions

In order to regulate the utilization and administration of water, the residents of San Gabriel Chilac have devised general regulations and maintain a record of agreements. The General Regulations, although not notarized before a notary public, were established by the users themselves since the inception of the branch model and have undergone modifications over time.

This regulatory framework delineates the structure and integration of the committees, outlines the interactions among all stakeholders, and specifies the authorities, responsibilities, limitations, as well as the rights and obligations of each party involved. It also outlines the formulation and execution of agreements. Additionally, the minutes of the agreements document the resolutions reached during the assemblies, including any sanctions imposed.

Both the Central Committee (CC) and the 52 Branch Committees (BCs) maintain their own regulations and memoranda of agreement. While the Regulations and Minutes of Internal Agreements for each BC adhere to the provisions outlined in the

General Regulations of the CC, these regulatory instruments can exhibit significant variations, particularly in the nature of agreements and the prescribed sanctions. This underscores the distinctive character and adaptability of each branch.

It is crucial to note that even though most of the agreements originate from the CC assembly, the final approval or disapproval rests with the users. An interesting aspect of local democracy is the requirement that members of the CC board of directors do not impose sanctions without prior consultation and authorization from the users. As expressed in the following testimony:

> If the CC detects a leak in a user's home, it must first make a recommendation to the user. Subsequently, it communicates the observation to the president of the branch so that the user assembly determines the fine (Interview with Member C, 2016).

Another noteworthy aspect is that 98.7% of users are aware of the specifications outlined in both the General Regulations and the agreements and sanctions established in their branch. Interviewed users mentioned that they acquired this information either by attending the assemblies or through communication from a family member who participated in these social gatherings. Since the agreements are established through consensus, users willingly adopt and adhere to them.

Certain provisions outlined in the Committee Regulations are rooted in local customs, such as the annual cooperation to cover the expenses of the patron saint's day and the mandatory inclusion of all users in the two committees, among other practices.

Broadly, the agreements and sanctions delineated in the regulations and minutes of agreements of both the CC and the BC are designed to ensure the responsible use of water and adherence to the agreements and obligations by all members. Some of the provisions and sanctions established in the CC regulatory framework include:

- The CC is authorized to impose a fine on the branch, rather than the branch president, if the president fails to attend the CC Assemblies.
- A branch failing to fulfill cleaning duties for water distribution channels is subject to a fine of 500 Mexican pesos (approx. 29.50 US dollars).
- The CC may fine a branch if it detects a user misusing water.
- Users pouring water into the drain are liable for 5000 pesos (approx. 295 US dollars) fine.
- Married children cohabiting in their parents' house are obligated to contribute 20% or 30% of the total cooperation for the cleaning activities of the conduits.

Some of the agreements and sanctions specified in the Internal Regulations of the branches are as follows:

- The President of the branch and the users comprising it are authorized to suspend a water intake to users who do not grant their corresponding cooperation or who refuse to hold any position in both the CC and the BC. The CC has no capacity to do so.
- The BC has the power to sanction the CC when it misappropriates resources and does not effectively fulfill its functions.

- The BC has the power to sanction the user who is absent unjustifiably (between 30 and 50 pesos or 1.75 and 2.95 US dollars fine). If the holder of the water intake cannot attend the Assembly, he must seek a person over 18 years of age to represent him.
- A user who does not participate in the maintenance activities of the water distribution system must pay a fine of $150 (approximately 9 US dollars).
- Failure to repair a water leak warrants a penalty of between 300 and 1000 pesos (between approx. 18 and 60 US dollars).
- Users who irrigate the patio with clean water must pay 1000 pesos fine (approx. 60 US dollars).

They have also implemented agreements that denote the presence of environmental awareness among users:

> Users of our branch agree to water the plants in the morning or at night and be careful not to waste water while we water them. We also agree not to waste water when we bathe (Interview with Member A, 2016).

> All the users of my town, we agreed to sweep the street every Sunday (Interview with Member D, 2016).

Unlike the case studies documented by Aguilar et al. (2001), the users of San Gabriel, in addition to establishing agreements regarding the proper use of water, have also implemented agreements that contribute to strengthening social cohesion in the community:

> If a user dies on our branch, we cooperate 50 pesos (approx. 2,95 US dollars) to support the family of the deceased (Interview with Member E, 2016).

> In my branch, we use the money from the fines to buy lights, patch the street and change the burned-out lights (Interview with Member F, 2016).

9.10 Roles and Responsibilities in the Water Governance Model

The Central Committee (CC) plays a crucial role in the efficient and fair distribution of water to the 52 branches within the management model. Its functions, as outlined in the regulations, encompass both technical and administrative aspects, and include the following:

(a) Maintain a constant volume of water in the storage tanks.
(b) Notify users of the scheduled days for water distribution rotations.
(c) Ensure cleanliness of the "La Taza" spring and distribution channels, free from debris and grass. Additionally, inspect storage tanks for any leaks.
(d) Conduct surprise inspection tours monthly to ensure users are not wasting water.

(e) Chlorinate the water and daily check the chlorine content in a water sample throughout the town.
(f) Approve new water intakes.
(g) Enforce the regulations' stipulations.
(h) Safeguard documentation, including official letters and assembly minutes, as well as the funds collected from fees.

9.11 Branch Committee

The BC has the following specific functions:

(a) Coordinate users to participate in cleaning and maintenance activities of the drinking water supply system, conducted semiannually.
(b) Collect and deliver users' financial contributions to the CC.
(c) Conduct a surprise inspection tour once a month to verify that branch users are not wasting water.
(d) Elevate problems that cannot be resolved at the branch level to the CC for determination of a solution.
(e) Safeguard documentation (official letters and assembly minutes) and funds collected from fees.
(f) Enforce the stipulations of the internal regulations.

To fulfill these general responsibilities, clear hierarchies are outlined in each committee, with specific duties delegated to each member.

9.12 Learning by Doing: Informal Knowledge Transmission

In contrast to some water management models where there is a formal obligation for members to pass on their roles, the committees in San Gabriel Chilac rely on a unique approach. Unlike the case study by Pimentel-Equihua et al. (2012), members of the Central Committee (CC) and the Branch Committee (BC) are not mandated to train their successors. Instead, members acquire the necessary skills and knowledge "on the fly," as articulated by the current president of the CC in an interview.

Despite the absence of formal training, the learning process is dynamic and enriched through shared knowledge among users and committee members. This informal knowledge transmission fosters a collaborative environment where insights and experiences are exchanged, contributing to the collective wisdom of the community in managing their water resources.

> There have been times when we present a problem to the Assembly and we do not know the way forward, some people who have an idea of how to solve it come to us to guide us (Interview with Member C, 2016).

Like many communities across the country, the studied community faces a lack of comprehensive data and information regarding critical aspects of their water management. Variables such as drinking water quality, available water quantity, water demand for various uses, and other technical, social, and financial information remain largely undocumented.

This absence of data presents a challenge to informed decision-making and planning for the sustainable management of water resources. Addressing this gap in information could potentially enhance the community's ability to devise more effective water strategies and policies, ensuring a more comprehensive understanding of their local water dynamics.

9.13 Accountability, Monitoring, and Follow-Up Mechanisms

In the intricate panorama of water governance, San Gabriel Chilac has implemented a robust set of accountability, monitoring, and follow-up mechanisms to ensure the transparent and effective management of water resources. These mechanisms are enshrined in the regulatory and normative framework, laying the foundation for responsible decision-making and sustained community involvement.

> The presidents of each committee must render accounts to the users on the way in which they implement the agreements, the results obtained and the achievement of their obligations. In the event that they do not perform their functions effectively and incur in diversion of resources, users have the power to sanction them (Interview with Member C, 2016).

They also make it easy to watch over what users are doing, checking if they're taking up committee roles, joining in cleaning efforts, and keeping up with contributions and fees. If someone falls short on their responsibilities, there are clear consequences in place. This system ensures that everyone is accountable for their actions, fostering a sense of responsibility among the community.

> A user who refuses to hold a position in any of the committees is sanctioned by the Users' Assembly for 5,000 pesos (approx. 295 US dollars). If he does not cover his debt, his water intake will be closed (Interview with Member G, 2016).

The CC must transparently present its financial performance during the Assemblies, subjecting itself to the evaluation of the users:

> Every three months, the president of the CC presents a report to the users about the money collected and the expenses incurred. In the assembly, the central president grants a copy of the report to the presidents of the 52 branches, so that, later, in their respective assemblies, they expose it to the users of their branch. If the users express any disagreement about the expenses incurred, the presidents of the branch must communicate it to the CC. This is how money is managed (Interview with Member C, 2016).

In a similar manner, the 52 branch committees provide a comprehensive report at the conclusion of their management period.

> Before the term of the branch committee ends, it must inform the assembly of how it managed the money collected, the expenses incurred and the existing capital (Interview with Member G, 2016).

Moreover, they are obligated to ensure that their assigned section of the water conduction channel is clean, free from grass and garbage.

> Before the end of the BC's management period, they are asked to clean up the section of the canal that they are supposed to clean because when the CC delivers the charges to the new CC, both committees inspect the entire section. If the branch does not comply with what was agreed, the CC will impose a sanction (Interview with Member C, 2016).

9.14 Financial Independence and Self-Financing Capacity

One of the challenges faced by water systems across various cities, municipalities, and localities in the country is the high rate of delinquency in service payments, leading to budget deficits (Pimentel-Equihua et al., 2012). However, in the town of Chilac, the recovery of fees and sanctions is 100%, as affirmed by the current president of the CC.

A unique aspect of this management model, contrasting with case studies reported by Aguilar et al. (2001), Pimentel-Equihua et al. (2012), and Sandoval Moreno and Günther (2015), is that the users of San Gabriel Chilac don't pay a monthly fee for the drinking water service; instead, they contribute sporadically (whenever there is an issue in the water infrastructure) and regularly (twice a year) to ensure the optimal functioning of the gravity-fed drinking water supply system and thereby maintain a reliable drinking water service. This is affirmed by the statements of two users:

> We take care of the maintenance of the water infrastructure. We cooperate with works or in cash to carry out repairs, change the pipe, pay for labor to clean the canal, etc. (Interview with Member D, 2016).

> When the branch president informs us that he needs to pay two or three people to clean the gutter, we cooperate to cover the payment (Interview with Member H, 2016).

In addition to sporadic contributions, users annually contribute to two fixed payments of 20 pesos (approx. 1,18 US dollars). The first occurs in September and is allocated for administrative expenses of the CC (e.g., payment for copies, transportation of chlorine, electricity used to inject chlorine into storage tanks, etc.). The second installment is collected in February to support the expenses of the patron saint's day.

The collection policy of this model also involves charging 50% interest on the total amount. This measure is implemented when the BC or any user fails to pay the agreed share within the established time. The funds generated from these payments (annual and sporadic) and from the sanctions imposed on the BC are managed by the CC. On the other hand, the BC manages the funds from fines imposed on users within its branch, as well as the contributions designated for the maintenance of the pipeline distributing water in its branch.

It's noteworthy that all members within this model are obligated to contribute as stipulated, including members of the CC and the BC, who do not receive compensation for their roles. This emphasis on social equity in fee contributions strengthens the trust between the CC and the users.

As evident, the devised collection and recovery mechanisms by the users ensure the payment of established contributions and sanctions, thereby enhancing the self-financing capacity and sustainability of the gravity-fed drinking water supply system.

9.15 Fair Water Distribution

For a considerable period until 2010, the residents of San Gabriel Chilac stood out from other municipalities in the subbasin by having access to drinking water every day of the year. However, in the post-2010 period, during the dry season (March to November), the demand for water from various users (agricultural, domestic, livestock) increased. To address this, the CC implemented a tandem distribution mode. Consequently, from March to November, each branch receives drinking water every 4 days a week, 24 h a day.

The water carrier distributes water based on the number of users within each branch. The quantity of water supplied to the 52 branches is measured by the number of turns made by the water carrier on the stopcock. The CC does not have information on the volume of water distributed each time the gate valve is opened.

According to the president of the CC, the coverage of the drinking water service in Chilac was reported as 100% in 2016. This claim was corroborated during fieldwork, as all 60 respondents affirmed having access to this service. At the community level, official data supports this, indicating that 98.7% of households were connected to piped water. Regarding water quality, 78% of respondents rated it as good, citing clear water that did not pose health risks. Another 21% classified it as regular due to perceived excess chlorine, while 1% considered it poor for human consumption due to inadequate quality.

Concerning the quantity of water, 83% of respondents mentioned that during the dry season, the water they receive only meets their basic needs but is insufficient for watering plants or patios. Conversely, 12% believed that they receive enough water, while 5% considered the amount to be inadequate. Among the 87% of users who reported regular or insufficient water supply, 98% purchase additional water in jugs, and 99% practice water reuse and adopt careful water usage practices, as illustrated in the table below.

With the provided information, it is evident that 28% of the sampled users engage in more than one water reuse practice or a combination of reuse, conservation, and efficient water use practices (Table 9.1).

Table 9.1 Percentage of the sampled population of San Gabriel Chilac that carries out water reuse, saving, and use activities

Activity of reuse, saving, and careful use of water	%
1	54
2	1
3	12
4	4
1–2	3
1–3	21
1–3–4	5
Total	**100**

Source: Own elaboration with data obtained in field work conducted by Griselda Tomé Hernández
1 = Reuse of used water (rinsing clothes) for household cleaning, sanitation, irrigation, etc
2 = Install water-saving devices
3 = The water with which you rinse the dishes is reused to wash dishes and water the patio
4 = Rainwater harvesting
(n = 60)

9.16 Conclusion

The case of San Gabriel Chilac serves as a notable example of community innovation and participation, showcasing how local actors have harnessed their physical resources and local capabilities to achieve a shared social objective: the collective management of drinking water. Leveraging the topography of the terrain, users implemented a gravity-driven water distribution system, thereby minimizing the need for electricity and demonstrating a sustainable approach to water management inherited from their precolonial and colonial predecessors.

Participatory, administrative, and collective mechanisms establish and shape a network-type management, with interaction among its members at the core of this model. The process of interaction unfolds in the assemblies, serving as a space for dialogue where all system members actively engage in communication, fostering a bidirectional flow from top to bottom and vice versa. The communication process for decision-making initiates in the CC, followed by dissemination through the BC, acting as the bridge for communication between the management team and the user base.

This network structure ensures the active participation of all members in the planning, execution, surveillance, and monitoring stages of agreements, actions, and functions. Simultaneously, it streamlines decision-making processes and facilitates the prompt and direct dissemination of information to users. Furthermore, it contributes to the early identification and resolution of issues, strengthening the bonds of trust within the community.

The structure, decision-making mechanisms, and components of the San Gabriel innovative model for drinking water allocation empower local actors to manage this vital resource independently of governmental oversight. It embodies various characteristics of participatory water governance, including a governing structure for water administration, distribution, surveillance, and monitoring; the creation of

Fig. 9.5 Photograph capturing the entrance of a man-made "galeria filtrante" or filtrating gallery. (Credit: Gravel, 2015).

spaces for consensus-driven decision-making; involvement of stakeholders in agreement formulation and implementation; clear definition and assignment of roles and responsibilities; regulatory frameworks; accountability; self-financing capacity; and features of social equity. Notably, the indicator of effective training is not observed in this model.

To advance participatory water governance, it is essential to recognize and reinforce the participatory capacities and knowledge of local actors. Additionally, analyzing and assessing their strategies and operational expertise in managing supply systems is crucial. Understanding the context dependency of governance is vital to effectively integrate water policies into diverse territories.

The specific case of drinking water management in the region has evolved from the convergence of at least two historic irrigation systems: the pre-Columbian system and the Spanish Castilian system. Central Mexico has a rich history of water provisioning strategies, including aqueducts with fountains in central city places (plazas) and boroughs, which are now part of colonial heritage assets. In the Tehuacán Valley, aside from dams and channels from natural springs (*manančiales*), filtrating galleries were also utilized to channel groundwater for irrigation, contributing to the agricultural intensification of the valley (Enge & Whiteford, 1989).

This chapter provides a glimpse into a significant period of Mexican agricultural history, acknowledging that it covers only a fraction of the vast narrative. For a more in-depth exploration, it is recommended to refer to the extensive literature on the subject as identified in the following bibliography (Fig. 9.5).

References

Aguilar, J., Gomez, T., Illsley, C., Flores, A., Quintanar, E., Tlacotempa, A., Acosta, J., & Mancilla, S. (2001). *Normas comunitarias indígenas y campesinas para el uso de los recursos naturales. Una experiencia de Guerrero, México*. Colección Manejo Campesino de Recursos Naturales. Grupo de estudios ambientales, AC. México, D.F. Retrieved at: https://gaiaoax.org/wp-content/uploads/2022/08/Normas-comunitarias-indigenas-y-campesinas.pdf. Accessed 14 Jan 2024.

Akhmouch, A. (2015). *Perspectivas de la OCDE en Gobernanza del Agua. Cuencas de Mexico*, n°. 2, año I, pp. 3–9.

Bakker, K. (2003). *Good governance in restructuring water supply: A handbook*. Federation of Canadian Municipalities.

Barriga, M., Campos, J. J., Corrales, O. M., & Prins, C. (2007). *Gobernanza ambiental, adaptativa y colaborativa en bosques modelo, cuencas hidrográficas y corredores biológicos Diez experiencias en cinco países latinoamericanos* (Informe técnico no. 358). Centro Agronómico Tropical de Investigación y Enseñanza, CATIE. Retrieved at: https://repositorio.catie.ac.cr/bitstream/handle/11554/1649/Gobernanza_ambiental_adaptativa_y_colaborativa.pdf?sequence=1&isAllowed=y. Accessed 14 Jan 2024.

Beaurain, C. (2002). Gouvernance environnementale locale et comportements économiques. *Développement durable et territoires, dossier, 2*, 1–15.

Brandes, O. M. (2005) At a Watershed: Ecological Governance and Sustainable Water Management in Canada. *Journal of Environmental Law and Practice*, vol. 16, n°. 1, pp. 79-97. Retrieved at: https://poliswaterproject.org/polis-research-publication/watershed-ecological-governance-sustainable-water-management-canada-2/. Accessed 14 Jan 2024.

Brody, A. (2009). *Género y gobernanza. Informe general, BRIGDE*. Retrieved at: https://america-latinagenera.org/wp-content/uploads/2014/03/genero_gobernanza_bridge2009.pdf. Accessed 14 Jan 2024.

Brooks, D. B., Brandes, O. M. and Gurman, S. (eds). (2009, 2011). *Making the most of the water we have: The soft path approach to water management*. : Earthscan.

Caire, G. (2004). Retos para la gestión ambiental de la cuenca Lerma Chapala: obstáculos institucionales para la introducción del manejo integral de cuencas. In A. H. Cotler (compiladora.) *El manejo integral de cuencas en México: estudios y reflexiones para orientar la política ambiental* (1st ed) (pp. 183–200). SEMARNAT-INE.

Cisneros, E. X. (2008). La gestión del agua a través de los cotas en México. Análisis de su gestión en cuatro estudios de caso. In D. Soares et al. (Eds.), *La gestión de los recursos hídricos: realidades y perspectivas* (First Vol) (pp. 257–280). Instituto Mexicano de Tecnología del Agua.

CONAPO (Consejo Nacional de Población). (2020). *Índices de marginación 2020*. Government of Mexico. Retrieved at: https://www.gob.mx/conapo/documentos/indices-de-marginacion-2020-284372. Accessed 14 Jan 2024.

De Boer, W. F., and Baquete, D. S. (1998) "Natural resource use, crop damage and attitudes of rural people in the vicinity of the Maputo Elephant Reserve, Mozambique". *Environmental Conservation, vol. 25*, n° 3, pp. 208-218.

Enge, K. I., & Whiteford, S. (1989). *The keepers of water and earth. Mexican rural social organization and irrigation*. University of Texas Press. Retrieved at: https://www.academia.edu/64356088/Globalization_Water_and_Health_Resource_Management_in_Times_of_Scarcity. Accessed 14 Jan 2024.

Enjolras, B. (2005). Économie sociale et solidaire et régimes de gouvernance". *RECMA, n° 296, pp. 56–69. Accessed Sept. 2, 2024.* https://www.erudit.org/fr/revues/recma/2005-n296-recma01112/1021862ar.pdf

Gagnon, C. (2000). Gouvernance environnementale et évaluation des impacts sociaux: un défi de citoyenneté. In Laurin, S. Klein, J.-L. and C. Tardif (dir.). Géographie et société. Sainte-Foy. Presses du l'Université du Québec, pp. 257–269.

Gil Huerta, G. (1972). History of the foundation of the town of San Gabriel Chilacatla. In R. S. MacNeish & F. Johnson (Eds.), *The Prehistory of the Tehuacán Valley, vol. 4, Chronology and Irrigation* (pp. 154–161). University of Texas Press.

Gleick, H. P. (2002) "Soft water paths". *Nature*, n° 418, pp. 373-376.

Government of Mexico. (2022). *Informe anual sobre la situación de pobreza y rezago social 2022*. Puebla, San Gabriel Chilac. Retrieved at: https://www.gob.mx/cms/uploads/attachment/file/698229/21_124_PUE_San_Gabriel_Chilac.pdf. Accessed 14 Jan 2024.

Government of the State of Puebla. (2014). *Programa Regional de Desarrollo 2011–2017*. Región Tehuacán y Sierra Negra.

Gravel, N., & Lavoie, A. (2009). Introduction: La gouvernance en perspective. *Revue canadienne des études latino-américaines et caraïbes, 34*(68), 12–18. Retrieved at: https://www.jstor.org/stable/i40084749. Accessed 14 Jan 2024.

Hernández Garciadiego, R., & Herrerías Guerra, G. (2004). Evolución de la tecnología hidro-agro-ecológica mesoamericana desde su origen prehistórico. Revista de Teología y Ciencias Humanas, vol. 69, n° 743. 14-19. Accessed August 30, 2024. http://www.alternativas.org.mx/Evolucion%20de%20la%20tecnologia.pdf

Heinen, J. T. (1996) "Human behavior, incentives and protected area management". *Conservation Biology*, vol. 10, n°. 2, 136-144.

INEGI. (2010). *Compendio de información geográfica municipal de los Estados Unidos Mexicanos*. San Gabriel Chilac.

INEGI (Instituto Nacional de Estadística y Geografía). (2020). *Censo de población y vivienda del Estado de Puebla*. Tabulados del cuestionario ampliado.

Infesta Dominguez, G., Vicente, A. and Cohen, I. (2012) "Reflexiones en torno al trabajo con grupos de discusión en ciencias sociales". Intersticios: Revista Sociológica de Pensamiento Crítico, vol. 6, n°. 1, pp. 233-244. Retrieved at: https://intersticios.es/article/view/8899. Accessed 14 Jan 2024.

Kauffer Michel, E. F. (2005). El consejo de cuenca de los ríos Usumacinta y Grijalva: los retos para concretar la participacion y la perspectiva de cuencas. In S. Vargas & É. Mollard (Eds.), *Problemas socioambientales y experiencias organizativas en las cuencas de México* (pp. 195–215). Jiutepec, Morelos. Retrieved at: https://horizon.documentation.ird.fr/exl-doc/pleins_textes/divers12-05/010038079.pdf. Accessed 14 Jan 2024.

Kauffer Michel, E. F. (2008). Comités de cuenca en Chiapas y Tabasco: entre participación, endeble y riesgo de politización. In D. Soares, S. Vargas, & M. R. Nuño (Eds.), *La gestión de los recursos hídricos: realidades y perspectivas* (Vol. I, pp. 193–227). Instituto Mexicano de Tecnología del Agua.

Moench, M., Dixit, A., Janakarajan, S., Rathore, M. S., & Mudrakartha, S. (2003). The Fluid Mosaic, Water Governance in the Context of Variability, Uncertainty and Change. Nepal Water Conservation Foundation, Kathmandu, and the Institute for Social and Environmental Transition, Boulder, Colorado. Accessd August 30, 2024 https://www.i-s-e-t.org/publications-and-resources-1/the-fluid-mosaic%3A-water-governance-in-the-context-ofvariability%2C-uncertainty-and-change

Mc C. Netting, R. (1981). *Balancing on an Alp: Ecological Change and Continuity in a Swiss Mountain Community. Cambridge: Cambridge University Press.*

Neely, J., Ajuvalasit, M., & Clause, V. (2015). New light on the prehistoric Purrón Dam Complex: Small corporate group collaboration in the Tehuacán Valley, Puebla, México. *Journal of Field Archaeology., 40*. https://doi.org/10.1179/2042458215Y.0000000010. Accessed 14 Jan 2024.

Ostrom, E. (2010). *Gouvernance des biens communs. Pour une nouvelle approche des ressources naturelles*. Bruxelles, De Boeck Supérieur. Col. « Planète en jeu ».

Pimentel Equihua, J. L. (2006). Organización social autogestiva y cogestión en el río Júcar: el caso de la acequia real del Júcar, comunidad de Valencia. Doctoral dissertation. https://doi.org/10.13140/RG.2.1.2445.9120

Pimentel-Equihua, J. L., Velazquez Machuca, M. A. and Palerm-Viqueira, J. (2012) "Capacidades locales y de gestión social para el abasto de agua doméstica en comunidades rurales del valle de Zamora, Michoacán, México". *Agricultura, Sociedad Y Desarrollo*, Vol. 9, n° 2, pp.107-121. Retrieved at: https://www.scielo.org.mx/scielo.php?script=sci_arttext&pid=S1870-54722012000200002. Accessed 14 Jan 2024.

Roulet, É. (2015). *L'évangélisation des Indiens du Mexique. Impact et réalité de la conquête spirituelle (XVIe c.).* Presses Universitaires de Rennes.

Sandoval Moreno, A. and Günther M. G. (2015) "Organización social y autogestión del agua Comunidades de la Ciénega de Chapala, Michoacán". *Política y Cultura n°*. 44, pp. 107-135.

Schneider, S., & Peyrè Tartaruga, I. G. (2006). Territorio y enfoque territorial: de las referencias cognitivas a los aportes aplicados al análisis de los procesos sociales rurales. In M. Manzanal, G. Neiman, & M. Lattuada (Eds.), *Desarrollo Rural. Organizaciones, Instituciones y Territorio* (pp. 71–102). Ed. Ciccus.

Soares, D., Vargas, S., & Nuno, M. R. (2009). Las percepciones sobre el agua en la cuenca del Río Amacuzac. In S. Vargas Velasquez & D. Soares Moraes (Eds.), *La gestión de los recursos hídricos: realidades y perspectivas* (Vol. 2, pp. 216–237). Instituto Mexicano de Tecnología del Agua.

Tomé Hernández, G. (2019). ≤Qué innovaciones sociales para la gestión participativa e integrada del agua en Río Salado, Puebla (México)? Análisis de la gestión integrada del agua de una subcuenca [Unpublished Ph.D. Dissertation, Université Laval]. http://hdl.handle.net/20.500.11794/36439

Theys, J. (2002) "La Gouvernance, entre innovation et impuissance". *Développement durable et territoires*, n° 2, pp. 1-28.

United Nations Commission on Global Governance. (1995). Our global neighbourhood. The report of the Commission on Global Governance. New York et Toronto : Oxford University. Accessed Sept. 2, 2024. https://www.gdrc.org/u-gov/global-neighbourhood/

UNESCO, The United Nations Educational, Scientific and Cultural Organization. (2003). *L'eau pour les hommes, l'eau pour la vie, Rapport mondial sur la mise en valeur des ressources en eau*. UNESCO/Division des sciences de l'eau.

UNESCO (2023). Imminent risk of a global water crisis warns the UN World Water Development Report 2023. Accessed Sept. 2, 2024. https://www.unesco.org/en/articles/imminent-risk-global-water-crisis-warns-unworld-water-development-report-2023

Vallejo, M., Ramirez, I. M., Reyes-Gonzalez, A., Lopez-Sanchez, J. G., & Casas, A. (2019). Agroforestry systems of the Tehuacan-Cuicatlan valley: Land-use for biocultural diversity conservation. *Land, 8*, 24. Retrieved at: https://doi.org/10.3390/land8020024. Accessed 14 Jan 2024.

Vargas Valencia, J. C., Diaz Nigenda, J. J., & Vargas Martinez, L. (2004). La gestión integrada de los recursos hídricos en México: Nuevo paradigma en el manejo del agua. In H. Cotler (comp), *El manejo integral de cuencas en México. Estudios y reflexiones para orientar la política ambiental*. Secretaría de Medio Ambiente y Recursos Naturales e Instituto Nacional de Ecología. México. Retrieved at: https://vortice.conagua.gob.mx/storage/files.conagua/upload/05022021_1612548107.pdf. Accessed 14 Jan 2024.

Vera Carta, J. (2005). Participación. Consejos de cuenca y política hidráulica mexicana: el caso de la costa de Chiapas. In S. Vargas & É. Mollard (Eds.), *Problemas socioambientales y experiencias organizativas en las cuencas de México* (pp. 276–297). IRD-IMTA. Retrieved at: https://horizon.documentation.ird.fr/exl-doc/pleins_textes/divers12-05/010038079.pdf. Accessed 14 Jan 2024.

Woodbury, R. B., & Neely, J.-A. (1972). Water control systems of the Tehuacan Valley. In R. S. MacNeish & F. Johnson (Eds.), *The prehistory of the Tehuacán Valley, vol. 4, chronology and irrigation* (pp. 81–153). University of Texas Press.

Chapter 10
Conclusion: Untangling the Knots in Environmental Governance

Abstract This chapter reflects on a decade of collaborative knowledge creation in environmental governance, emphasizing intercultural dialogues and integrating diverse perspectives. It highlights the importance of inclusive governance frameworks that recognize and integrate scientific and local knowledge. Case studies across the Americas illustrate innovative approaches in blue, green, and bee city planning and community participation, underscoring the role of grassroots initiatives in sustainable development. The chapter advocates for adaptive management practices and meaningful dialogue to address environmental challenges collaboratively.

Keywords Environmental governance · Collaborative knowledge creation · Intercultural dialogues · Inclusive governance frameworks · Scientific and local knowledge · Green city planning · Community participation · Grassroots initiatives · Sustainable development · Adaptive management practicesBlue/Green/Bee framework

The journey depicted encompasses both personal and professional dimensions, emphasizing the collaborative knowledge creation with graduate students and research collaborators over a decade. This collaboration included diverse individuals and scholars encountered during field research. Intercultural dialogues played a crucial role, prompting me to revisit and deepen my understanding of key concepts. These experiences and insights have been integrated into the context of this book on environmental citizenship and participatory governance, giving new meaning to past studies.

The co-constructed knowledge, shared through words or actions, has bridged various points and illuminated intersubjective and intercultural boundaries, highlighting the importance of respecting differences and recognizing sensitivities. Identifying points of friction and diverging perceptions have refined my skills as a geographer, providing a nuanced roadmap that includes the consequences and uncertainties of different encounters and paths. Completing this book represents an embrace of uncertainty and a commitment to navigating and living with it.

N. Gravel, *Tending Nature*, https://doi.org/10.1007/978-3-030-47062-3_10

Several paths have been identified for learning communities and decision-makers who possess the power to transform our world. It is imperative for power-holders to recognize that their knowledge is but a fragment of the larger body of existing knowledge. While this knowledge may be validated by scientific standards, it coexists with other forms of knowledge from diverse sources, which are equally valid.

In conducting case studies across North, Central, and South American and the Caribbeans contexts, the clash of cultures has prompted significant questions for both myself and my collaborators. Embracing the Zapatista philosophy of "walking while questioning" has been essential. The lessons from these intercultural exchanges are extensive but can only be partially captured in words. Following the conclusions of Briones and Ramos (2021: 27), who worked with the Mapuche in Chile, I acknowledge that the experiential knowledge underlying this research is richer and more nuanced than what can be conveyed here, presenting a substantial challenge in articulation. Here are insights gained from intercultural sharing (Idem):

- Reciprocal teaching and learning
- Collaboration with knowledge holders
- Challenging established certainties
- Navigating diverse ontologies and epistemologies
- Equality and rights in knowledge co-construction
- Significance of conversation beyond information exchange[1]
- Recognizing boundaries and frontiers of interculturality or interdisciplinarity
- Acknowledging differences and tensions while maintaining open communication channels.

On another level, when communities come together to act, aiming to achieve consensus in formulating a response or collaboratively planning for their future, they evolve into learning communities (Prévost & Sévigny, 2005). Lessons gleaned from studying such communities reveal that certain prerequisites must be met before engaging in discussions about the fair and respectful management of natural resources (Gravel, 2023): The preparatory stages for community learning are diverse and should be tailored to each community's unique needs. Capacity building, access to education, and relevant information are often essential for communities to feel empowered and participate effectively in resource management (Gravel and Martinez Arboleya, 2015). This preparation involves formal and informal education, open-minded discussions, appreciation of local knowledge, and access to technology and reliable scientific monitoring results.

Adhering to the principles of adaptive management, it is crucial to adopt a cautious approach, allowing for necessary adjustments and technological adaptations (Meek, 2010: 205). Active monitoring and evaluation of new technologies should precede the progression to the next stage of development, as Meek (Idem: 206)

[1] "The importance of conversation lies in establishing new connections with what remains disjointed within us" according to Briones and Ramos (2021: 27).

succinctly puts it, "Move forward while we continue to improve knowledge." The early stages of the journey toward community governance are summarized below. The state's role in this process is integral to providing a conducive environment for the empowerment of civic groups (Idem; Gravel, 2023).

Here is a synthesis of preparatory stages that a community may envision for a more inclusive governance:

- Identify geographical scales corresponding to socio-environmental issues
- Promote the multifunctionality of rural spaces through government policies
- Foster the formation of learning communities
- Improve access to information and education to enhance stakeholders' capacities
- Establish partnerships and create support networks
- Implement natural resource management regulatory processes (with the Government).

As mentioned earlier, it is crucial to consider the geographical boundaries of the territory and the jurisdiction of various actors, ranging from the local level affected by conflicting uses to the administrative scales involved in decision-making. In order to have an impact on decisions, community representatives must secure their presence and an equitable voice in negotiations at the highest levels of management. A comprehensive understanding of how the phenomenon is perceived at different scales and the consistent presence in each of these scales are essential to ensure the successful inclusion of groups that are typically marginalized in negotiations (Zimmerer, 2006; Reid et al., 2006).

The case studies presented in this book, drawn from Canada, the United States, Mexico, Costa Rica, Brazil, and partially from Cuba, showcase some of the most forward-thinking innovations in participatory environmental governance in the Americas. These examples contribute to the development of what we synthesize as the Blue/Green/Bee framework for conservation, encompassing blue and green city planning, participatory natural resource management (water and forest), nature and bee-scaping, urban and peri-urban agriculture, foodshed citizenship, and the conservation of native pollinator diversity.

Environmental stewardship can be passed down from one generation to the next or learned. In urban environments, the rupture with nature occurred generations ago, and local knowledge about nature-tending has been lost or is threatened. However, some dedicated gardeners have maintained the art of cultivating in urban settings (discussed in Chap. 8). Gardeners interested by emerging agroecological practices should start by acquiring fundamental knowledge about companion planting, seed selection, native plants, soil regenerative techniques, native bee conservation, tree-crop coexistence, and waste management through composting. The Brazilian urban gardening network exemplifies a knowledge co-construction community. It plays a pivotal role on the web, facilitating the sharing and construction of practical knowledge among practitioners, addressing their queries effectively. Group workshops, collaborative tasks, and activist demonstrations in solidarity with fellow urban gardeners forge the bonds among the green guerrilleros.

Grassroots organizations and smart-city staff as well reach out to the public, raise awareness, and disseminate information. Public programs created by innovative city planners, such as those of Guelph, Kitchener-Waterloo, and the region of Peel in southern Ontario, Canada, promote environmental landscaping (naturescaping) and water-wise practices. These programs encourage citizens to adopt sustainable practices in their outdoor water use. Benefits include water savings, the rediscovery of native plants adapted to local conditions, the attraction of local wildlife, low-maintenance yards, savings on lawn treatments, and the collection of water runoffs using permeable surface materials. Rainwater barrels are also made available to citizens.

The success of these municipal policies raises the question of how these water-wise landscaping techniques were conveyed to the population. The answer lies in outreach to residents through handouts, publicity at fairs, local newspapers, and city web pages, as well as workshops, demonstrative gardens, and free consultations. The geographic spread of innovation within a neighborhood, driven by imitation, appears to amplify concepts like naturescaping or xeriscaping. These ideas are so compelling that even those who may be hesitant can be enticed to convert a portion of their lawn into a rock garden or a garden featuring native plants due to their low maintenance appeal.

Eco-networking, the result of the work of ecological web weavers (as discussed in Chaps. 3, 4, 5, 6, and 8), using the Internet to facilitate information exchange and partnerships between private, public, semipublic entities, and civic organizations, has emerged as a result of creative thinking, solidarity needs, and communication webs to reach a common objective. One additional consequence arising from the anticipation of climate change is the establishment of comprehensive platforms that unite individuals, organizations, governments, and businesses, ensuring community resilience amid the challenges of rising temperatures and diminished rainfall. This looming threat is and has been real enough to get different stakeholders to sit around the table and start planning for the future. The example of the region of Peel in Southern Ontario is telling (Chap. 5): Six governmental entities, comprising the region of Peel and three municipalities, collaborated with two conservation authorities to devise climate change strategies for the period 2011–2050. Their comprehensive report assessed various aspects, including water conservation, water runoff management, urban forests, alternative energy technologies, construction norms for new buildings, and the urban transportation system. This thorough evaluation led to a reexamination and consolidation of existing green techniques and water-smart strategies, all framed within the overarching strategy of climate change mitigation.

In that sense, the catalytic power of the looming climate change acts to structure environmental conservation actions and mobilize citizens in a way that would not have occurred without facing the threat of warmer summers, water shortages, and devastating flash floods. Add to that the common water source shared with the city of Toronto, Lake Ontario, and the threats posed by its receding coastline and biodiversity, and you've got sufficient reasonable reasons to persuade a great number of citizens to start caring for their surrounding environment, for the water creek they had never noticed before, the marshes that were also ignored, and that big factory that wants to expand its activities along a stream. As residents of the Peel region became more engaged, they started financing conservation authorities and participating in meetings and activities, spreading the word around them. The result is dynamic well-structured organizations embedded in their communities that attract the attention of nationals from everywhere, one of them being the Credit River Conservation authority. Another consequence of getting the wheel turning is a gain in influence at the political level and the attraction of experts to sit on organization boards. The success is local and is starting to radiate at a broader geographical scale.

There are numerous stories of local populations' involvement in decision-making processes in Latin America and elsewhere. The management of the Reventazón Model Forest in Costa Rica (Chap. 4) and the decentralized water management innovation of San Gabriel Chilac in Mexico (Chap. 9) appear to be success stories worth learning from. In these chapters, experiences of natural resource and territorial co-management have been reviewed and documented from the perspective of the actors who work on a daily basis. Despite various projects being carried out with the help of international funding agencies to bring the benefits of forest conservation, the International Model Forest Network offers a credible alternative that is more inclusive and representative of local populations in Costa Rica and elsewhere.

Among the most spectacular successes to celebrate are the reversal of the Ontarian PM's decision to remove the Greenbelt's protection on lands surrounding the Greater Toronto Area (Chap. 3) and the demolition of dams on the main stems of salmon rivers in Maine (Chap. 6). These achievements result from a considerable number of actors working together within eco-networks, bringing together uncommon allies, sometimes even former opponents, such as fishing clubs and First Nations. Mapping these actors allows us to see their diversity of backgrounds and trajectories, identifying those who were crucial in convincing the rest of the population to embark on their projects and pressure local governments. The media played an important role in these two stories, acting as a whistleblower in Ontario (Chap. 3) and a consolidating agent for the river revitalization movement in Maine (Chap. 6).

The safeguarding of wild native pollinators in rural and urban mosaics is clearly more than a biodiversity conservation issue; it is paramount to the preservation of the human species and the delicate equilibrium that still exists. The loss of pollination services would deprive us humans of more than 35% of existing edible plants and would deal a blow to the agri-food system and its actors. Many actors in production, transformation, transportation, and sales would be impacted, along with small local businesses that have flourished and contribute to diversifying the local foodstuff offer. Industries such as apples and blueberries would be among the first ones impacted, representing food staples in many cultures, not to mention a source of vitamins in northern countries where local fruit alternatives are limited. Agroecology, agroforestry, and ecologically intensive agriculture are favorable avenues for conserving diversified rural landscapes capable of providing pollen reservoirs for wild pollinators. More incentives should be given by public decision-makers to encourage small organic/natural productions that respect principles of ecological reconciliation, moving away from monoculture and sterilized landscapes toward transition landscapes.

The proposed transition landscapes outlined here align with the Blue-Green-Bee Framework, a holistic approach where all components are interconnected, akin to river ecosystems (Chap. 6). This book advocates not only for municipalities and regional governments to not only adopt the Blue-Green-Bee Framework as an objective but also to embrace it as the path forward in designing transition landscapes that deviate from the crisis-driven trajectory we have been following. The journey toward ecological reconciliation necessitates embracing a stewardship approach rooted in the residents' actions within their foodsheds or watersheds. Continuous awareness and critical questioning among citizens and society regarding their values, choices, consumption habits, and waste generation and disposal, form the foundation of meaningful personal and collective engagement in the health of their ecosystems. It is imperative to recognize that human well-being is intimately connected to the health of the environments they inhabit, and there is no substitute for planet Earth. How about becoming an engaged citizen of your food/watershed today?

References

Briones, C., & Ramos, A. M. (2021). Revitalizing conversations: Lessons from and about the production of intersubjective and intercultural knowledge. *The Canadian Journal of Action Research, 21*(3), 17–31. Retrieved at: https://journals.nipissingu.ca/index.php/cjar/issue/view/40

Gravel, N., & Martinez Arboleya, H. J. (2015). A Social Partnership for Peacebuilding in Mexico: Examples from the States of Nuevo León and Sonora. In the special issue ed. by N. Gravel on The Return of Hope in Mexico: The Implementation of Peace-Building Processes in a Tensed Social Environment. Journal of the Canadian Association of Latin American and Caribbeans Studies, 40 (2): 176-199. Accessed Sept. 2, 2024. http://www.tandfonline.com/doi/abs/10.1080/08263663.2015.1044719?journalCode=rclc20#preview

Gravel, N. (2023). Antecedentes y trayectorias de la gobernanza participativa de los recursos naturales: lecciones de Canadá, Costa Rica y Brasil. In G. Massardier & V. Ugalde (Eds.), *Analizar la nueva fábrica de las políticas públicas: pluralización, territorialización, participación* (pp. 471–516). Colegio de México. ISBN 978-607-564-562-94.

Meek, S. (2010). Integrated watershed planning approaches for addressing uncertainty in urbanizing areas. In B. Mitchell (Ed.), *Resource and environmental management in Canada: Addressing conflict and uncertainty* (4th ed., pp. 204–206). Oxford University Press.

Prévost, P., & Sévigny, B. (2005). La collectivité apprenante comme stratégie de développement local. *The Journal of Community Informatics.* June issue. Retrieved at: http://vision2025.uqac.ca//capitalcreativite/prevostsevigny.pdf

Reid, W. V., Berkes, F., Wilbanks, T. J., & Capistrano, D. (Eds.). (2006). *Bridging scales and knowledge systems. Concepts and applications in ecosystem management.* Island Press.

Zimmerer, K. S. (2006). *Globalization and new geographies of conservation.* The University of Chicago Press.

Index

N. Gravel, *Tending Nature*, https://doi.org/10.1007/978-3-030-47062-3

The manufacturer's authorised representative in the EU is Springer Nature Customer Service Centre GmbH, Europaplatz 3, 69115 Heidelberg, Germany. If you have any concerns regarding our products, please contact ProductSafety@springernature.com

Printed and bound by CPI Group (UK) Ltd, Croydon, CR0 4YY
01/10/2026
02225962-0009